Lecture Notes in Mathematics

Edited by A. Dold and B. Eckmann

707

Roland Zielke

Discontinuous Čebyšev Systems

Springer-Verlag
Berlin Heidelberg New York 1979

Author
Roland Zielke
Universität Osnabrück
Fachbereich 5
Albrechtstraße 28
D-4500 Osnabrück

Library of Congress Cataloging in Publication Data

Zielke, Roland, 1946-
Discontinuous Čebyšev systems.

(Lecture notes in mathematics ; 707)
Bibliography: p.
Includes index.
1. Chebyshev systems. I. Title. II. Series:
Lectures notes in mathematics (Berlin) ; 707.
QA3.L28 no. 707 [QA355] 510'8s [511'.4] 79-12233

AMS Subject Classifications (1970): 26A51, 34C10, 41A30, 41A50

ISBN 3-540-09125-4 Springer-Verlag Berlin Heidelberg New York
ISBN 0-387-09125-4 Springer-Verlag New York Heidelberg Berlin

Printing and binding: Beltz Offsetdruck, Hemsbach/Bergstr.
2141/3140-543210

Preface

While Čebyšev systems basically are part of classical analysis, they gain their importance in various fields of applications such as stochastics or approximation theory. After the publication of KARLIN's and STUDDEN's comprehensive monograph on continuous Čebyšev systems (1966), in recent years there has been increasing interest in the discontinuous case as well. The aim of this book is to present some of these developments in a coherent and unified manner.

We have restricted ourselves to fundamental properties of Čebyšev systems. For example, we did not try to develop a theory of moment spaces of discontinuous functions though this might be a promising subject of future research. Also excluded is the investigation of splines or generalized splines being weak Čebyšev systems; it seems likely, however, that some of the methods developed here could be useful in this area. In dealing with generalized convex functions we have only formulated some of the more important results and omitted various ramifications and special cases that certainly are of independent interest.

We only deal with real-valued Čebyšev systems(one aspect of the complex-valued case is referred to in the notes for chapter 2).

There are few overlappings with existing books; most of the material has until now been available only in journals. About half of the proofs are new and in many instances simpler than those given in the original papers. This is partly due to the heavy use of alternations instead of determinants. While some theorems are fairly straightforward generalizations of corresponding results for the continuous case, some of the material seems to be substantially new (results 6.5, 8.8, 10.5, 12.2, 23.3; see the notes).

The notes and references are far from exhaustive. Following the reference lists of the papers and books cited, however, it should not be difficult to get a fairly complete view of the earlier literature relevant to the subject.

To most chapters a few exercises are appended. Some of them comprise results referred to in subsequent chapters. Part of the exercises may require some work; these are marked by an asterix "$*$".

In the field of Čebyšev systems no unified terminology exists; terms like "T-space", "Tschebyscheff system", "Haar space", "T-system" etc. are used with different meanings by different authors. Our choice of terminology - though aimed at brevity and simplicity - is necessarily

subjective. Anyway - "Name ist Schall und Rauch".
Open problems are stated on p. 7, 33, 41, 43, 46, 48, 74, 80, 88, 93, 99, 103.

I would like to thank Mrs. A. Schmidt for her careful typing of the manuscript.

Table of contents

Notations

M = domain of definition

$\mathbb{R}$ = real numbers

C = complex plane

$S^1 = \{z \in C \mid |z| = 1\}$

$F = \{f: M \to \mathbb{R}\}$

$B(M) = \{f \in F \mid f \text{ bounded}\}$

$C(M) = \{f \in F \mid f \text{ continuous}\}$

$C_+(M) = \{f \in F \mid f \text{ continuous from the right}\}$

$C_-(M) = \{f \in F \mid f \text{ continuous from the left}\}$

$\Delta_n(A) = \{(x_1,\dots,x_n) \in A^n \mid x_1 < \dots < x_n\}$ for $A \subset \mathbb{R}$

$\bar{\Delta}_n(A) = \{(x_1,\dots,x_n) \in A^n \mid x_1 \leq \dots \leq x_n\}$ for $A \subset \mathbb{R}$

$\bar{\Delta}_n^{(r)}(A) = \{(x_1,\dots,x_n) \in \bar{\Delta}_n(A) \mid x_1 < x_{i+r}$ for $i = 1,\dots,n-r\}$

$\# A$ = cardinality of A.

$Z(f)$ = set of all zeros of an $f \in F$

$SZ(f)$ = set of all simple zeros of an $f \in F$ } (see p. 24)

$DZ(f)$ = set of all double zeros of an $f \in F$ }

I. Domains of definition

1. Generalized Haar spaces

Let M be a non-empty set and F the linear space of all real-valued functions defined on M.
For a subset A of M and a subset U of F let $U|_A$ be the set of functions that are restrictions of elements of U to A.
Usually we shall be dealing with subsets $U \subset F$ that are finite-dimensional linear spaces with certain additional properties.
If $U \subset F$ is an n-dimenional linear space, we denote by U^* the set of all linear functionals on U, and by $U^*_M \subset U^*$ the set of all point functionals, i. e., x^* is in U^*_M if and only if there is an $x \in M$ such that for all $f \in U$ we have $x^*(f) = f(x)$.
We omit the proof of the following lemma which can be found in textbooks on linear algebra.

Lemma 1.1: Let $U \subset F$ be an n-dimensional linear space, $n \geq 1$. Then there is a subset N of M consisting of n points such that $U|_N$ is an n-dimensional linear space.

We are now ready for the first basic definition:

Definition: Let $U \subset F$ be an n-dimensional linear space, $n \geq 1$. U is called a generalized Haar space (on M, with respect to M) if every function in U has at most $n - 1$ zeros or vanishes identically. A basis of a generalized Haar space is called a generalized Čebyšev system.

Apart from this definition of generalized Haar spaces via the zeros, other equivalent formulations are available. Some of these are listed in the next result:

Lemma 1.2: Let $U \subset F$ be an n-dimensional linear space, $n \geq 1$. Then the following properties are equivalent:

a) U is a generalized Haar space.

b) For every subset N of M consisting of n elements, $U|_N$ is an n-dimensional generalized Haar space.

c) Each set of n pairwise distinct point functionals $x^*_1, \ldots, x^*_n \in U^*_M$ is linearly independent.

d) For every subset $\{x_1, \ldots, x_n\}$ of M consisting of n points and for every basis $f_1, \ldots, f_n$ of U, we have $\det(f_i(x_j)) \neq 0$.

e) For every subset $\{x_1, \ldots, x_n\}$ of M consisting of n points and arbitrary real numbers $\alpha_1, \ldots, \alpha_n$ there is exactly one $f \in U$ with $f(x_j) = \alpha_j$ for $j = 1, \ldots. n$.

Proof: The equivalence of a) and b) is obvious. Now let $f_1,\dots, f_n$ be a basis of U.

a) is equivalent to the statement a') that the system of linear equations

$$\sum_{i=1}^{n} \gamma_i f_i(x_j) = 0, \quad j = 1,\dots, n,$$

has only the trivial solution. This is the case iff the matrix $(f_i(x_j))_{n,n}$ is regular, i. e., d) holds.

Moreover, a') is equivalent to e).

Statement d) means that the column vectors of the matrix $(f_i(x_j))_{n,n}$ are linearly independent. As $f_1,\dots, f_n$ form a basis of U, this is equivalent to the linear independence of the point functionals $x_1^*,\dots, x_n^*$, i. e., c).

From lemma 1.2 one easily derives the following result which will turn out to be quite useful on various occasions:

Lemma 1.3: Let U be an n-dimensional generalized Haar space on M, $n \geq 2$, and $x_1,\dots, x_k \in M$ be k pairwise distinct points, $k \leq n - 1$. Then the restriction of $V := \{f \in U \mid f(x_1)=\dots=f(x_k)=0\}$ to the set $\tilde{M} := M \setminus \{x_1,\dots,x_k\}$ is an (n-k)-dimensional generalized Haar space on $\tilde{M}$.

Proof: Clearly no nonvanishing element of $V|_{\tilde{M}}$ has more than $n - 1 - k$ zeros. As to the dimension of $V|_{\tilde{M}}$, let $x_{k+1},\dots, x_n \in \tilde{M}$ be pairwise distinct points and $g_1,\dots, g_n \in U$ be defined by

$$g_i(x_j) = \delta_{i,j} \quad \text{for } i,j = 1,\dots, n.$$

Then $g_{k+1}|_{\tilde{M}},\dots, g_n|_{\tilde{M}}$ are linearly independent and span $V|_{\tilde{M}}$, for otherwise $g_1,\dots, g_n$ would be no basis of U.

Lemma 1.4: Let U be an n-dimensional generalized Haar space on M, $n \geq 2$, N a set and ϕ a mapping from N into M. Then $\tilde{U} := \{h: N \to \mathbb{R} \mid h = f\circ\phi \text{ for an } f \in U\}$ is an n-dimensional generalized Haar space on N iff ϕ is injective.

Proof: If an $h \in \tilde{U} \setminus \{0\}$ had n zeros $x_1,\dots, x_n$, the function $f \in U$ with $h = f \circ \phi$ would have n zeros $\phi(x_1),\dots, \phi(x_n)$, and ϕ would not be injective.

On the other hand, if we had $\phi(x) = \phi(y)$ for some $x, y \in N$ with $x \neq y$, the point functionals $x^*, y^* \in \tilde{U}_N^*$ would be linearly dependent and by lemma 1.2 c) $\tilde{U}$ would not be a generalized Haar space.

<u>Corollary 1.5:</u> Let U be an n-dimensional generalized Haar space on M, $n \geq 2$, N a set and $\psi : M \to N$ an injective mapping. Then $\hat{U} := \{h:\psi(M)\to\mathbb{R} \mid h=f\circ\psi^{-1} \text{ for an } f\in U\}$ is an n-dimensional generalized Haar space.

<u>Lemma 1.6:</u> Let U be an n-dimensional generalized Haar space on M, $n \geq 2$. Then there exists an injective mapping $\psi : M \to \mathbb{R}^n$.

<u>Proof:</u> Let $f_1,\ldots, f_n$ be a basis of U. The mapping $\psi : M \to \mathbb{R}^n$ defined by

$$\psi(x) = (f_1(x),\ldots,f_n(x)) \quad \text{for } x \in M$$

is injective because of lemma 1.2 c).

As $\mathbb{R}$ and $\mathbb{R}^n$ are of the same cardinality, the last result immediately yields:

<u>Theorem 1.7:</u> Let U be an n-dimensional generalized Haar space on M, $n \geq 2$. Then there exists an injective mapping $\psi : M \to \mathbb{R}$.

Instead of the above reasoning, a more explicit argument for theorem 1.7 may be given:

<u>Proof:</u> Let $x_1,\ldots, x_n \in M$ be pairwise disjoint points. By lemma 1.3 there exists a generalized Čebyšev system f, g on $M \setminus \{x_1,\ldots,x_{n-2}\}$. We may assume $f(x_{n-1}) = 0$ because of lemma 1.2. Now either $\tilde{M}:=M \setminus \{x_1,\ldots,x_{n-1}\}$ consists of one point, and the statement of theorem 1.7 is trivial, or f and g form a generalized Čebyšev system on $\tilde{M}$, and f does not vanish on $\tilde{M}$. In this case, if we denote the constant function by 1, 1 and $\frac{g}{f}$ again form a generalized Čebyšev system on $\tilde{M}$. By lemma 1.2 c), $\frac{g}{f}$ is injective on $\tilde{M}$, and so there is an injective mapping from M into $\mathbb{R}$, as well.

<u>Exercises:</u> 1) Let $M = \mathbb{R}$ and $f_i(x) = x^{i-1}$ for $x \in M$, $i = 1,\ldots, n$. Show that $f_1,\ldots, f_n$ form a generalized Čebyšev system on $\mathbb{R}$.
2) Let M be a set of at least $k \geq 1$ points, and assume there is an injective mapping $\psi : M \to \mathbb{R}$. Show that generalized Haar spaces of arbitrary dimension $n \leq k$ may be defined on M.

2. Domains of definition with a topology

In this chapter we shall consider a more special situation than in the preceding one. We make the additional assumption that M is endowed with a topology and that the functions under consideration are continuous. The linear space of continuous functions on M is denoted by $C(M) \subset F$. From lemma 1.4 immediately follows:

Lemma 2.1: Let M and N be topological spaces, $U \subset C(M)$ an n-dimensional generalized Haar space on M, and $\phi : N \to M$ an injective, continuous mapping. Then $\tilde{U} := \{h \in C(N) \mid h = f \circ \phi$ for some $f \in U\}$ is an n-dimensional generalized Haar space.

Lemma 2.2: Let M be a topological space and $U \subset C(M)$ an n-dimensional generalized Haar space, $n \geq 2$. Then M is a Hausdorff space.

Proof: The mapping $\psi : U \to \mathbb{R}^n$ that was defined in the proof of lemma 1.6, is continuous and injective. As $\mathbb{R}^n$ is Hausdorff, the same holds for M.

If, in addition to the hypotheses of lemma 2.2, M is compact, ψ is a homeomorphism from U into $\mathbb{R}^n$. However, the following sharper results hold:

Theorem 2.3: a) Under the hypotheses of lemma 2.2, there is a continuous, injective mapping from M into the unit circle.
b) If, in addition, M is compact, M is homeomorphic to a subset of the unit circle.

While b) follows immediately from a), for the proof of a) we need various auxiliary results most of which are topological in essence. We recall that a topological space $\underline{\bar{X}}$ is called disconnected if there is a nontrivial separation of $\underline{\bar{X}}$, i. e., two nonempty, open (closed) subsets A, B of $\underline{\bar{X}}$ with $\underline{\bar{X}} = A \cup B$ and $A \cap B = \emptyset$. Otherwise $\underline{\bar{X}}$ is called connected.

Proposition A: Let $\underline{\bar{X}}$ be a connected topological space, and for some $x \in \underline{\bar{X}}$ let A and B constitute a nontrivial separation of $\underline{\bar{X}} \setminus \{x\}$. Then x is a boundary point of A and of B, and $A \cup \{x\}$ and $B \cup \{x\}$ are connected.

Proof: As $\underline{X}$ is connected, we have $\bar{A} \cap \bar{B} \neq \emptyset$. So $\bar{A} \cap \bar{B} = \{x\}$, and x is a boundary point of A as well as of B. Now suppose $\bar{A} = A \cup \{x\}$ is disconnected, and let A_1 and A_2 be a nontrivial separation of $\bar{A}$ with $x \in A_2$. But then A_1 and $A_2 \cup B$ are a nontrivial separation of $\underline{X}$.

Proposition B: Let $\underline{X}$ be a topological space, and assume that every mapping from $\underline{X}$ into the unit circle $S^1 \subset \mathbb{R}^2$ that is continuous and injective, is onto. Let ϕ be such a mapping. Then $\phi^{-1}(P)$ is connected if $P \subset S^1$ is connected.

Proof: 1) $\underline{X}$ is connected. Suppose the contrary, i. e., $\underline{X} = A \cup B$ is a nontrivial separation of $\underline{X}$. Then $\phi(A)$ and $\phi(B)$ are homeomorphic to subsets of $\mathbb{R}$, and so $\phi(A \cup B)$ is homeomorphic to a subset of $\mathbb{R}$, too. Hence there is a continuous, injective, but not bijective mapping from $\underline{X}$ into S^1.
2) Suppose for some $x \in \underline{X}$, A and B are a nontrivial separation of $\underline{X} \setminus \{x\}$. By proposition A, each of the sets $A \cup \{x\}$, $B \cup \{x\}$ is connected. Since ϕ is continuous, $\phi(A \cup \{x\})$ and $\phi(B \cup \{x\})$ are connected subsets of S^1 with $\phi(x)$ as the only common point. In the same fashion as in the first part of the proof, we may then construct a continuous, injective, but not bijective mapping from $\underline{X}$ into S^1 and arrive at a contradiction.
3) Now let P be a connected subset of S^1 with endpoints a and b and $a \neq b$. Let $\underline{Y} := P \setminus \{a,b\}$ and $Z := S^1 \setminus (\underline{Y} \cup \{a,b\})$. As $\underline{Y}$ and Z are a nontrivial separation of $S^1 \setminus \{a,b\}$, $\phi^{-1}(\underline{Y})$ and $\phi^{-1}(Z)$ form a nontrivial separation of $\underline{X} \setminus \{\phi^{-1}(a), \phi^{-1}(b)\}$. Because of b), $\underline{X} \setminus \{\phi^{-1}(a)\}$ and $\underline{X} \setminus \{\phi^{-1}(b)\}$ are connected. By proposition A each of the sets $\phi^{-1}(\underline{Y} \cup \{a\})$, $\phi^{-1}(\underline{Y} \cup \{b\})$, $\phi^{-1}(Z \cup \{a\})$, $\phi^{-1}(Z \cup \{b\})$ is connected. So $\phi^{-1}(\underline{Y} \cup \{a,b\})$ is also connected. Finally, if $\phi^{-1}(\underline{Y})$ were disconnected we would again get a contradiction, using proposition A. In any case $\phi^{-1}(P)$ is connected.

Lemma 2.4: Let the hypotheses of proposition B be fulfilled, and assume $C(\underline{X})$ contains an n-dimensional generalized Haar space U. Then n is odd.

Proof: Let $p_1, \ldots, p_n \in S^1$ with $0 \leq \arg p_1 < \ldots < \arg p_n < 2\pi$, and $p_{n+1} := p_1$. Let P_i be the closed arc from p_i to p_{i+1}, $i = 1, \ldots, n$. Let $x_i := \phi^{-1}(p_i)$ and $\underline{X}_i := \phi^{-1}(P_i)$, $i = 1, \ldots, n$. By proposition B, each $\underline{X}_i$ is connected. Now let $f \in U$ be a function with $f(x_i) = (-1)^i$ for $i = 1, \ldots, n$, in accordance with lemma 1.2 e), the interpolation property of generalized Haar spaces. As f is continuous, each set $f(\underline{X}_i)$ is connected. If n were even, each set $f(\underline{\overline{X}}_i)$, $i = 1,\dots,n$, would contain the points -1 and +1 and thus a zero of f. f would then have n zeros.

Proposition C: Let $\underline{\overline{X}}$ be a topological space, and assume that every proper subspace of $\underline{\overline{X}}$ can continuously and injectively be mapped into $\mathbb{R}$. Then $\underline{\overline{X}}$ can continuously and injectively be mapped into S^1.

Proof: 1) Let $\underline{\overline{X}} = A \cup B$ be a nontrivial separation. As A and B can be mapped continuously and injectively into $\mathbb{R}$, the statement is obvious.
2) Let $\underline{\overline{X}}$ be connected, and assume that for an $x \in \underline{\overline{X}}$, the set $\underline{\overline{X}} \setminus \{x\}$ has a nontrivial separation into A and B. Let $b \in B$, and $\phi : \underline{\overline{X}} \setminus \{b\} \to \mathbb{R}$ be an injective, continuous mapping. By proposition A, $\phi(A \cup \{x\})$ is an interval with $\phi(x)$ as a boundary point. Analogously, an injective, continuous mapping can be found that maps $B \cup \{x\}$ into an interval with $\phi(x)$ as a boundary point. From these two mappings one easily constructs a mapping from $\underline{\overline{X}}$ into $\mathbb{R}$ which is injective and continuous.
3) Let $\underline{\overline{X}}$ be connected, and for any $p \in \underline{\overline{X}}$ let $\underline{\overline{X}} \setminus \{p\}$ be connected. Take an $x \in \underline{\overline{X}}$ and an injective and continuous mapping $\phi : \underline{\overline{X}} \setminus \{x\} \to \mathbb{R}$. So $\phi(\underline{\overline{X}} \setminus \{x\})$ is an interval. Let $y \in \underline{\overline{X}} \setminus \{x\}$ be chosen such that $\phi(\underline{\overline{X}} \setminus \{x,y\})$ is disconnected, so $\underline{\overline{X}} \setminus \{x,y\}$ is disconnected. Let $\underline{\overline{X}} \setminus \{x,y\} = A \cup B$ be a nontrivial separation. As $\underline{\overline{X}} \setminus \{x\}$ and $\underline{\overline{X}} \setminus \{y\}$ are connected, by proposition A each of the sets $A \cup \{x\}$, $B \cup \{x\}$, $A \cup \{y\}$, $B \cup \{y\}$ is connected, and so are $A \cup \{x,y\}$ and $B \cup \{x,y\}$. Moreover, x and y are boundary points of A and B. $A \cup \{x,y\}$ and $B \cup \{x,y\}$ can continuously and injectively be mapped onto closed intervals with endpoints corresponding to x and y. So there is a continuous, bijective mapping from $\underline{\overline{X}}$ onto S^1.

Proof of theorem 2 a): n = 2: For a basis f, g of U, a continuous, injective mapping $h : M \to S^1$ is defined by

$$h(x) = \left(\frac{f(x)}{\sqrt{f^2(x)+g^2(x)}}, \frac{g(x)}{\sqrt{f^2(x)+g^2(x)}} \right) \quad \text{for } x \in M.$$

n - 1 => n: If M consists of n points, the statement is trivial. If M contains more than n points, $U|_{M \setminus \{x\}} \subset C(M \setminus \{x\})$ is an n-dimensional generalized Haar space for every $x \in M$. Besides, if we set $\tilde{U} := \{f \in U \mid f(x) = 0\}$, $\tilde{U}|_{M \setminus \{x\}}$ is an (n-1)-dimensional generalized Haar space in $C(M \setminus \{x\})$. By the induction hypothesis there is a continuous, injective mapping from $M \setminus \{x\}$ into S^1. Because of lemma 2.4 there must exist another such mapping that, in addition, is not onto. So $M \setminus \{x\}$ can continuously and injectively be mapped into $\mathbb{R}$ for every $x \in M$.

Finally apply proposition C.

A direct specialization of lemma 2.4 is

Theorem 2.5: Every generalized Haar space in $C(S^1)$ has odd dimension.

We mention without proof that the statement of theorem 2.3 b) remains true if one replaces the compactness of M by the hypothesis that M is locally compact and connected, or that M is locally compact and the elements of the Haar space "vanish at infinity" (see LUTTS 1964). Topological characterizations of those sets M that can be topologically imbedded into S^1 if C(M) contains a generalized Haar space of dimension ≥ 2, do not seem to be known.

Exercises: 1) Fill in the details of the proof of lemma 2.2.
2)* If one wishes to prove only theorem 2.3 b), some of the auxiliary results are superfluous or may be replaced by simpler arguments. Find a short, self-contained proof of theorem 2.3 b). (The references given in the notes of this chapter may be helpful).

3. Totally ordered domains of definition, Haar spaces

The results of the last chapter suggest that many generalized Haar spaces of interest will be defined on subsets of the real line. Leaving topological aspects aside for the moment, we shall more generally consider sets M that are endowed with a total order "<", i. e., for every pair $x, y \in M$ exactly one of the following statements is true: $x < y$ or $y < x$ or $x = y$.
We begin with some notations:
Let $\Delta_n(M) = \{(x_1,\ldots,x_n) \in M^n \mid x_1 < \ldots < x_n\}$.
For an $f \in F$, k points $x_1, \ldots, x_k \in M$ are called a strong (weak) alternation of f of length k iff the following conditions hold:
a) $x_1 < \ldots < x_k$, and
b) $(-1)^i f(x_i)$ is positive (nonnegative) for all i, or $(-1)^i f(x_i)$ is negative (nonpositive) for all i.
The central concept of this book is contained in the following

Definition: An n-dimensional generalized Haar space $U \subset F$, $n \geq 1$, is called a Haar space (on M) iff no $f \in U$ has a strong alternation of length n + 1. A basis of a Haar space is called a Čebyšev system.

In the following lemma we give two equivalent formulations of the concept of Haar spaces that are basic to most considerations throughout the rest of this book.

<u>Lemma 3.1:</u> Let $U \subset F$ be an n-dimensional linear space, $n \geq 1$. Then the following properties are equivalent:

a) U is a Haar space.

b) No $f \in U \setminus \{0\}$ has a weak alternation of length $n + 1$.

c) Let $f_1,\ldots,f_n \in U$ be a basis. Then $\det\,(f_i(x_j))_{n,n}$ has constant sign for all $(x_1,\ldots,x_n) \in \Delta_n(M)$.

<u>Remark:</u> From lemma 1.1 and lemma 1.2 it is obvious that the sign of the determinant in lemma 3.1 is not identically zero.

<u>Proof of lemma 3.1:</u> If M consists of n points all the statements a), b) and c) hold. So we now assume $\Delta_{n+1}(M) \neq \emptyset$.

<u>c) => a):</u> The remark after lemma 3.1 shows, together with lemma 1.2, that U is a generalized Haar space. Without loss of generality suppose there is an $f \in U$ and $(x_1,\ldots,x_{n+1}) \in \Delta_{n+1}(M)$ with $(-1)^i f(x_i) < 0$ for $i = 1,\ldots,n+1$. From c) we get:

$$\det\begin{pmatrix} f & f_1 & \cdots & f_n \\ x_1 & \cdots\cdots & & x_{n+1} \end{pmatrix} := \begin{vmatrix} f(x_1) & f_1(x_1) & \cdots & f_n(x_1) \\ \vdots & \vdots & & \vdots \\ f(x_{n+1}) & f_1(x_{n+1}) & \cdots & f_n(x_{n+1}) \end{vmatrix} =$$

$$= \sum_{i=1}^{n+1} (-1)^{i-1} f(x_i) \det\begin{pmatrix} f_1 \cdots\cdots\cdots\cdots f_n \\ x_1 \ldots x_{i-1} x_{i+1} \ldots x_{n+1} \end{pmatrix} =$$

$$= \sum_{i=1}^{n+1} |f(x_i)| \det\begin{pmatrix} f_1 \cdots\cdots\cdots\cdots f_n \\ x_1 \ldots x_{i-1} x_{i+1} \ldots x_{n+1} \end{pmatrix} \neq 0.$$

But then U has dimension $n + 1$, a contradiction.

<u>a) => b):</u> Suppose there is an $f \in U \setminus \{0\}$ and $(x_1,\ldots,x_{n+1}) \in \Delta_{n+1}(M)$ with $(-1)^i f(x_i) \geq 0$ for $i = 1,\ldots,n+1$. As f has at most $n - 1$ zeros, we have $(-1)^k f(x_k) > 0$ for some $k \in \{1,\ldots,n+1\}$. Because of the interpolation property of generalized Haar spaces there exists (exactly one) $g \in U$ with $g(x_j) = (-1)^j$ for $j = 1,\ldots,k-1,k+1,\ldots,n+1$. For all $\alpha > 0$ and $j = 1,\ldots,k-1,k+1,\ldots,n+1$ we have $(-1)^j(f+\alpha g)(x_j) > 0$. For sufficiently small $\alpha > 0$, we also get $(-1)^k(f+\alpha g)(x_k) > 0$. So for some $\alpha > 0$, $f + \alpha g \in U$ has a strong alternation of length $n + 1$ in contradiction to the hypothesis.

b) ⇒ c): Let $f_1,\ldots,f_n$ be a basis of U, and without loss of generality let $(x_1,\ldots,x_n) \in \Delta_n(M)$ with $\det(f_i(x_j)) > 0$. Now suppose $\det(f_i(y_j)) \le 0$ for some $(y_1,\ldots,y_n) \in \Delta_n(M)$. Let $W: = \{x_1,\ldots,x_n\} \cup \{y_1,\ldots,y_n\}$, and $w_1,\ldots,w_n$ be the first n points of W. We form the sets

$$A_i: = \begin{cases} \{w_1,\ldots,w_{i-1},x_i,\ldots,x_n\} & \text{for } i = 1,\ldots,n+1 \\ \{w_1,\ldots,w_{2n+1-i},y_{2n+2-i},\ldots,y_n\} & \text{for } i = n+2,\ldots,2n+1. \end{cases}$$

Each set A_i consists of n points, and A_i differs from A_{i+1} in at most one point, $i = 1,\ldots,2n$. If we consider the determinant corresponding to each A_i, it becomes clear we only have to deal with the case $x_i=y_i$ for $i = 1,\ldots,k-1,k+1,\ldots,n$, $x_k \neq y_k$, for some fixed $k \in \{1,\ldots,n+1\}$. Let $g \in U$ be defined by

$$g(x): = \det\begin{pmatrix} f_1 \ldots\ldots\ldots\ldots\ldots\ldots f_n \\ x_1 \ldots x_{k-1}\; x\; x_{k+1} \ldots x_n \end{pmatrix} = \begin{vmatrix} f_1(x_1) \ldots\ldots f_n(x_1) \\ \vdots \\ f_1(x_{k-1}) \ldots f_n(x_{k-1}) \\ f_1(x) \ldots\ldots f_n(x) \\ f_1(x_{k+1}) \ldots f_n(x_{k+1}) \\ \vdots \\ f_1(x_n) \ldots\ldots f_n(x_n) \end{vmatrix}$$

g has zeros in $x_1,\ldots,x_{k-1},x_{k+1},\ldots,x_n$, and $g(x_k) > 0 \ge g(y_k)$. Then either $x_1,\ldots,x_k,y_k,x_{k+1},\ldots,x_n$ or $x_1,\ldots,x_{k-1},y_k,x_k,\ldots,x_n$ form a weak alternation of g of length n+1, a contradiction.

Lemma 3.2: Let $U \subset F$ be an n-dimensional Haar space, and $f \in F$ a positive (negative) function. Then
$\tilde{U}: = \{h \in F \mid h = f \cdot g \text{ for some } g \in U\}$
is an n-dimensional Haar space.

We shall now show that, in a sense, the consideration of arbitrary totally ordered sets M instead of $M \subset \mathbb{R}$, means but a pseudo-generalization. The following two theorems may be considered as analoga of theorems 1.7 and 2.3.

Theorem 3.3: Let U be an n-dimensional Haar space on M, $n \ge 2$. Then there is a strictly monotone mapping from M into $\mathbb{R}$.

Proof: For M finite, the statement is trivial. Otherwise let $q \in M$ be such that at least n-1 points of M lie to the left of q and to the right of q. Let $(x_1,\ldots,x_{n-1}) \in \Delta_{n-1}(M)$ with $x_{n-1} < q$,

$P := \{x \in M | x_{n-1} \leq x\}$, $Q := \{x \in M | q \leq x\}$. By lemmas 1.3 and 3.1, the restriction of

$$V := \{f \in U | f(x_1) = \ldots = f(x_{n-2}) = 0\}$$

to P is a two-dimensional Haar space.
Let f,g be a basis of V with $f(x_{n-1}) = 0$. By lemma 3.1 b), f has constant sign $\neq 0$ on Q. If 1 denotes the constant function, 1 and $\frac{g}{f}$ form a Čebyšev system on Q because of lemma 3.2. Invoking lemma 3.1 again, the determinant condition c) yields the strict monotonicity of $\frac{g}{f}$ on Q.
In an analogous manner a strictly monotone mapping from $\tilde{Q} := \{x \in M | x \leq q\}$ into $\mathbb{R}$ is found. A strictly monotone mapping of M into $\mathbb{R}$ can now be pieced together in an obvious way.

The total order of M in a natural way induces a topology on M: The topology generated by the system of subsets

$$T := \{A \subset M | A = \{p | p < x\} \text{ or } A = \{p | p > x\} \text{ or } A = \{p | x < p < y\} \text{ for some } x,y \in M\}$$

is called order topology on M. In the following let $C(M) \subset F$ be the set of all functions in F that are continuous with respect to this topology.

Theorem 3.4: Let $U \subset C(M)$ be an n-dimensional Haar space, $n \geq 2$. Then there is a strictly monotone , homeomorphic mapping from M into a subset of $\mathbb{R}$.

Proof: The mapping $\frac{g}{f}$ from the proof of theorem 3.3 is continuous on Q under the additional hypothesis of theorem 3.4. So a continuous, strictly monotone mapping $\phi: M \to \mathbb{R}$ can be constructed. Such a mapping, howewer, is a homeomorphism onto $\phi(M)$.

In most textbooks on approximation theory M is assumed to be a real interval, and the functions considered are continuous. In this case the terms "Haar space" and "generalized Haar space" coincide.

Theorem 3.5: Let M be a real interval and $U \subset C(M)$ an n-dimensional generalized Haar space. Then U is a Haar space.

Proof: If an $f \in U$ had a strong alternation of length n+1, f would have n zeros.

Exercises: 1) Let M be totally ordered and $U \subset F$ an n-dimensional Haar space. Let $f \in U \setminus \{0\}$ with n-1 zeros $x_1 < \ldots < x_{n-1}$, and $f_1, \ldots, f_n$ a basis of U.

a) Show that f is a scalar multiple of the function $g \in U$ that is defined by

$$g(x) := \det \begin{pmatrix} f_1 \ldots\ldots f_n \\ x\ x_1 \ldots x_{n-1} \end{pmatrix} \text{ for } x \in M.$$

b) Let $x_0 := \inf M$, $x_n := \sup M$, and $X_i := \{x \in M \mid x_{i-1} < x < x_i\}$, $i = 1,\ldots,n$. For all $p_i \in X_i$ and $p_j \in X_j$ one has $\operatorname{sign}(f(p_i) \cdot f(p_j)) = (-1)^{i-j}$, $i,j = 1,\ldots,n$, i.e., f changes sign exactly in its zeros.

2) Give a detailed argument for the last sentence in the proof of theorem 3.4.

3) a) Let $M \subset \mathbb{R}$, $U \subset F$ an n-dimensional linear space and $\varphi \in F$ be strictly monotonous. Then $U \subset F$ is a Haar space if and only if

$$\hat{U} := \{g\colon \varphi(M) \to \mathbb{R} \mid g = f \circ \varphi^{-1} \text{ for some } f \in U\}$$

is a Haar space.

b) In addition, let M be an interval and $\varphi \in C(M)$. Then one has $U \subset C(M)$ if and only if $\hat{U} \subset C(\varphi(M))$.

4)* Let $M = \{0,1,2,3\}$. Find a three-dimensional generalized Haar space U such that there is no bijective mapping $\varphi\colon M \to M$ transforming U into a Haar space $\tilde{U} = \{h \in F \mid h = f \circ \varphi \text{ for an } f \in U\}$ (see ZIELKE 1971).

II. Haar spaces and related concepts

In view of the results of chapter 3 we shall throughout the rest of this book only consider Haar spaces of functions defined on subsets of the real line, in other words, we shall generally assume $M \subset \mathbb{R}$.

4. Weak Haar spaces

In the last chapter the alternation property was introduced as an addition to the zero property of generalized Haar spaces. In this chapter we shall investigate alternation properties separately.
It turns out that the central statements of Lemma 3.1 are in fact formed by the synopsis of two independent characterizations, namely lemma 1.2 and the following result:

Lemma 4.1: Let $U \subset F$ be an n-dimensional linear space. Then the following statements are equivalent:
a) No $f \in U$ has a strong alternation of length $n + 1$.
b) If $f_1,\dots,f_n$ is a basis of U, $\det(f_i(x_j))$ has weakly constant sign on $\Delta_n(M)$ (i.e., nonnegative on $\Delta_n(M)$ or nonpositive on $\Delta_n(M)$).
c) For every $(u_1,\dots,u_{n-1}) \in \Delta_{n-1}(M)$ there is an $f \in U \setminus \{0\}$ with zeros in $u_1,\dots,u_{n-1}$ and $(-1)^i f(x) \geq 0$ for $x \in (u_{i-1},u_i) \cap M$, $i = 1,\dots,n$, where $u_0 := -\infty$ and $u_n := +\infty$.

We are thus led to the following

Definition: An n-dimensional linear space $U \subset F$ is called a weak Haar space if U has one (hence all) of the properties a), b), c) in lemma 4.1. A basis of a weak Haar space is called a weak Čebyšev system.

Proof of lemma 4.1: a) => b) and c) => b): Let $f_1,\dots,f_n$ be a basis of U. We suppose there are $(x_1,\dots,x_n)$, $(y_1,\dots,y_n) \in \Delta_n(M)$ with $\det(f_i(x_j)) \cdot \det(f_i(y_j)) < 0$.
As we have $\dim U|_{\{x_1,\dots,x_n\}} = \dim U|_{\{y_1,\dots,y_n\}} = n$, there exists a y_{i_1} with $\dim U|_{\{y_{i_1},x_2,\dots,x_n\}} = n$, then a y_{i_2} with $\dim U|_{\{y_{i_1},y_{i_2},x_3,\dots,x_n\}} = n$ and so on. So $\{x_1,\dots,x_n\}$ can - step by step - be transformed into $\{y_1,\dots,y_n\}$ such that the corresponding determinant does not vanish at any step. At least once the determinant must change sign, say at the k-th step.
For simplicity we rename $y_{i_1},\dots,y_{i_{k-1}},x_k,\dots,x_n$ into $z_1,\dots,z_n$ with $z_1<\dots<z_n$ and y_{i_k} into y. Let $z_j < y < z_{j+1}$ with $z_0 := -\infty$ and $z_{n+1} := \infty$, and let z_p be replaced by y. The determinant corresponding to any basis of U changes sign at this step. We define a basis $g_1,\dots,g_n$ of U by

$g_\mu(z_\nu) = \delta_{\mu,\nu}$ for all μ,ν.

Because of $\det\begin{pmatrix} g_1 \cdots\cdots g_n \\ z_1 \cdots\cdots z_n \end{pmatrix} = 1$ we get

$$0 > \begin{cases} \det\begin{pmatrix} g_1 \cdots\cdots\cdots\cdots\cdots\cdots\cdots\cdots g_n \\ z_1 \ldots z_j \; y \; z_{j+1} \ldots z_{p-1} z_{p+1} \ldots z_n \end{pmatrix} & \text{for } j = 0,1,\ldots,p-2,p+1,\ldots,n \\ \det\begin{pmatrix} g_1 \cdots\cdots\cdots\cdots g_n \\ z_1 \ldots z_{p-1} \; y \; z_{p+1} \ldots z_n \end{pmatrix} & \text{for } j = p-1,p. \end{cases}$$

Computation of these determinants gives

$$\operatorname{sign} g_p(y) = \begin{cases} (-1)^{p-j} & \text{for } j = 1,\ldots,p-2 \\ -1 & \text{for } j = p-1,p \\ (-1)^{p-j-1} & \text{for } j = p+1,\ldots,n. \end{cases}$$

So g_p has a weak alternation of length $n + 1$ in $z_1,\ldots,z_j,y,z_{j+1},\ldots,z_n$ and $n - 1$ zeros $z_1,\ldots,z_{p-1},z_{p+1},\ldots,z_n$.

So far, we have not used the hypotheses a) or c).

<u>If a) holds</u> we define - recalling that $U|\{z_1,\ldots,z_n\}$ is a Haar space - $g \in U$ by

$$g(z_m) = \begin{cases} (-1)^{j-1-m} \operatorname{sign} g_p(y) & \text{for } m = 1,\ldots,j \\ (-1)^{j-m} \operatorname{sign} g_p(y) & \text{for } m = j+1,\ldots,n. \end{cases}$$

One easily checks $g(z_p) = 1$. For sufficiently small $\varepsilon > 0$, $g_p + \varepsilon g \in U$ has a strong alternation of length $n + 1$ in $z_1,\ldots,z_j,y,z_{j+1},\ldots,z_n$ contradicting hypothesis a).

<u>If c) holds</u> there is an $f \in U \setminus \{0\}$ with zeros in $z_1,\ldots,z_{p-1},z_{p+1},\ldots,z_n$ with weak sign changes in each zero and weakly constant sign on each set $(z_i,z_{i+1}) \cap M$, $i = 0,\ldots,p-2,p+1,\ldots,n$ and on $(z_{p-1},z_{p+1}) \cap M$. As $U|\{z_1,\ldots,z_n\}$ is a Haar space f must be a scalar multiple of g_p (see ch.3, exercise 1). But then f has the "wrong" sign either in y or in z_p.

<u>b) ⇒ c):</u> Let $(u_1,\ldots,u_{n-1}) \in \Delta_{n-1}(M)$, $u_0 := -\infty$, $u_n := +\infty$. Let $f_1,\ldots,f_n$ be a basis of U and denote by φ the column vector $\begin{pmatrix} f_1 \\ \vdots \\ f_n \end{pmatrix}$.

We proceed by induction over $k := \dim U|\{u_1,\ldots,u_{n-1}\}$:

<u>k = n-1:</u> If we define $f \in U$ by

$$f(t) := \det\begin{pmatrix} f_1 \cdots\cdots f_n \\ t \quad u_1 \ldots u_{n-1} \end{pmatrix} \text{ for } t \in M,$$

f does not vanish identically because of $\dim U = n$. As we have

$$f(t) = (-1)^i \det \begin{pmatrix} f_1 \ldots\ldots\ldots\ldots\ldots f_n \\ u_1 \ldots u_i \; t \; u_{i+1} \ldots u_{n-1} \end{pmatrix}$$

for $t \in (u_i, u_{i+1}) \cap M$, $i = 0,1,\ldots,n-1$, hypothesis b) implies that f has the desired sign properties.

<u>$k+1 \Rightarrow k$:</u> Let $i_1,\ldots,i_k \in \{1,\ldots,n-1\}$ be such that $\varphi(u_{i_1}),\ldots,\varphi(u_{i_k})$ are linearly independent, $L := \{w \in M \mid \varphi(w) \in \operatorname{span}\{\varphi(u_{i_1}),\ldots,\varphi(u_{i_k})\}\}$, and for $\nu = 1,\ldots,n-1$ let L_ν be the largest interval with $u_\nu \in L_\nu \subset L$. We distinguish two cases:

<u>case 1:</u> There is a $\nu \in I := \{1,\ldots,n-1\} \setminus \{i_1,\ldots,i_k\}$ with $[u_{\nu-1}, u_{\nu+1}] \cap M \not\subset L_\nu$, say $W := (u_{\nu-1}, u_\nu) \cap M \setminus L \neq \emptyset$ without loss of generality.
If W has a maximal element x, by induction hypothesis there is an $f \in U \setminus \{0\}$ with weak sign changes in $u_1,\ldots,u_{\nu-1},x,u_{\nu+1},\ldots,u_{n-1}$. As f vanishes on L, f has the desired sign property.
If W has no maximal element, let $\{x_m\}_{m=1}^{\infty}$ be a strictly increasing sequence in W converging to sup W . By induction hypothesis for every $m = 1,2,\ldots$ there is an $h_m \in U$ with $\|h_m\| = 1$, $\|\;\|$ being some norm on U, and weak sign changes in $u_1,\ldots,u_{\nu-1},x_m,u_\nu,\ldots,u_{n-1}$.
As the unit sphere $S := \{f \in U \mid \|f\| = 1\}$ is compact with respect to the norm topology, the sequence $\{h_m\}$ contains a subsequence converging to some $f \in S$. One easily checks that f has the desired sign properties (norm convergence implies pointwise convergence; see also chapters 7 and 8).

<u>case 2:</u> For each $\nu \in I$ we have $[u_{\nu-1}, u_{\nu+1}] \cap M \subset L_\nu$. Now let $\nu \in I$ be fixed, and without loss of generality let $\mu < \nu$ be maximal such that $W := (u_\nu, u_{\nu+1}) \cap M \setminus L$ is not empty. The rest of the proof is exactly the same as in case 1, with the only modifications that if W has a maximal element, $f \in U \setminus \{0\}$ is constructed with weak sign changes in

$$u_1,\ldots,u_\mu,x,u_{\mu+1},\ldots,u_{\nu-1},u_{\nu+1},\ldots,u_{n-1},$$

and otherwise $h_m \in S$ is constructed with weak sign changes in

$u_1,\ldots,u_\mu,x_m,u_{\mu+1},\ldots,u_{\nu-1},u_{\nu+1},\ldots,u_{n-1}$, $m = 1,2,\ \ldots$.

<u>b) => a):</u> We suppose there is an $h \in U$ and $(x_1,\dots,x_{n+1}) \in \Delta_{n+1}(M)$ with $(-1)^i\, h(x_i) < 0$ for $i = 1,\dots,n+1$. Let $U_1 := U|_{\{x_1,\dots,x_{n+1}\}}$.
We distinguish two cases:
<u>Case 1:</u> $\dim U_1 = n$. Let $g_1,\dots,g_n$ be a basis of U_1. From

$$0 = \det\begin{pmatrix} h\ g_1 \dots g_n \\ x_1 \dots\dots x_{n+1}\end{pmatrix} = \sum_{i=1}^{n+1} (-1)^{i-1} h(x_i) \det\begin{pmatrix} g_1 \dots\dots\dots\dots\dots g_n \\ x_1 \dots x_{i-1} x_{i+1} \dots x_{n+1}\end{pmatrix}$$

follows

$$\det\begin{pmatrix} g_1 \dots\dots\dots\dots\dots g_n \\ x_1 \dots x_{i-1} x_{i+1} \dots x_{n+1}\end{pmatrix} = 0 \text{ for } i = 1,\dots,n+1.$$

But then $\dim U_1 < n$, a contradiction.
<u>Case 2:</u> $\dim U_1 := k < n$. We shall show that in this case another strong alternation of length $n+1$ may be constructed such that the restriction of U to the corresponding points has dimension $k+1$. This process may then be repeated until we reach case 1.
Let $g_1,\dots,g_n$ be a basis of U, and denote by ϕ the column vector $\begin{pmatrix} g_1 \\ \vdots \\ g_n\end{pmatrix}$, set

$$\phi_1 := \phi(x_1),\dots,\phi_{n+1} := \phi(x_{n+1}),$$
$$A := \{\phi_1,\dots,\phi_{n+1}\} \text{ and } V := \operatorname{span} A,$$

so $\dim V = k$. Let $t \in M$ be such that $\phi(t) \notin V$, so $\dim(\operatorname{span}\{V,\phi(t)\}) = k+1$. This implies there is an $f \in U$ with $f(x_1) = \dots = f(x_{n+1}) = 0$ and $f(t) = 1$, in other words, arbitrary values may be assigned to h in t without changing h on $\{x_1,\dots,x_{n+1}\}$.
Setting $x_0 := -\infty$ and $x_{n+2} := \infty$, we have $t \in (x_\nu, x_{\nu+1})$ for some ν.
If it exists, let x_μ be the first point right of t such that the set $A \setminus \{\phi_\mu\}$ contains a basis of V. So for every $p \in \{\nu+1,\dots,\mu-1\}$ the set $A \setminus \{\phi_p\}$ does not span V. For such a p, let $\phi_{i_1},\dots,\phi_{i_k}$ be a basis of V with $i_1 = p$. Let the basis $g_1,\dots,g_n$ of U be specified by $g_l(t_j) = \delta_{l,j}$, $l,j = 1,\dots,n$, where $t_j = x_{i_j}$ for $j = 1,\dots,k$ and $t_{j+1},\dots,t_n \in M$ are chosen such that $U|_{\{t_1,\dots,t_n\}}$ has dimension n.
Clearly $g_1,\dots,g_k$ is a basis of U_1. If we had $g_1(x_r) \neq 0$ for some $r \in \{2,\dots,n+1\}$, we would get

$$\phi_p = \frac{1}{g_1(x_r)}\left(\phi_r - \sum_{j=2}^{k} g_j(x_r)\cdot \phi_{i_j}\right),$$

and $\phi_r, \phi_{i_2},\dots,\phi_{i_k}$ would be a basis of V not involving ϕ_p, a contra-

diction. So we get $g_1(x_p) = 1$, $g_1|\{x_1,\ldots,x_{p-1},x_{p+1},\ldots,x_{n+1}\} = 0$. So arbitrary values may be assigned to h in each $x_p, p \in \{\nu+1,\ldots,\mu-1\}$ without changing h in $x_1,\ldots,x_\nu,x_\mu,\ldots,x_{n+1}$.
We set $h(t) := h(x_{\nu+1})$ and then reverse the sign of h in $x_{\nu+1},\ldots,x_{\mu-1}$. So h has a strong alternation of length n+1 in $x_1,\ldots,x_\nu,t,x_{\nu+1},\ldots,x_{\mu-1},\ldots,x_{n+1}$, and we have $\dim(\mathrm{span}\{\phi_1,\ldots,\phi_{\mu-1},\phi(t),\phi_{\mu+1},\ldots,\phi_{n+1}\}) = k+1$.
If there exists no point x_μ with the above property and $t < x_\mu$ there must be such an x_μ with $x_\mu < t$, and an analogous argument applies.

As a special case of lemma 4.1 we get:

Corollary 4.2: If $U \subset F$ is an n-dimensional weak Haar space, for every $(t_1,\ldots,t_n) \in \Delta_n(M)$ there is an $f \in U \setminus \{0\}$ with $(-1)^i f(t_i) \geq 0$ for $i = 1,\ldots,n$.

Points in M where all functions of a linear space $U \subset F$ vanish, clearly are of no importance for the alternation condition by which weak Haar spaces are defined. In the following we shall call such points unessential points of M (with respect to U), all other points essential points.
Furthermore we define: If $f \in F$ has zeros x,y with $x < y$, x and y are called separated zeros of f if there is a $z \in M$ with $x < z < y$ and $f(z) \neq 0$. With these notations we are now ready to analyze the structure of weak Haar spaces more closely.

Theorem 4.3: Let $U \subset F$ be an n-dimensional weak Haar space, and assume M consists of essential points only. Then
a) every $f \in U$ has at most n separated zeros, and
b) if an $f \in U$ has n separated zeros $x_1<\ldots<x_n$, f vanishes left of x_1 and right of x_n.

For the proof we need the following auxiliary result:

Proposition A: Let $U \subset F$ be an n-dimensional linear space, and assume M consists of essential points only. Then for every $(x_1,\ldots,x_n) \in \Delta_n(M)$ there is a $g \in U$ with $g(x_i) \neq 0$ for $i = 1,\ldots,n$.

<u>Proof:</u> We show that for each $k \le n$ there is a $g \in U$ with $g(x_i) \neq 0$ for $i = 1,\dots,k$. For $k = 1$ this is trivial.
<u>$k - 1 \Rightarrow k$:</u> Let $g \in U$ with $g(x_i) \neq 0$ for $i = 1,\dots,k-1$. If $g(x_k) \neq 0$, nothing has to be shown. Otherwise let $h \in U$ with $h(x_k) \neq 0$. For sufficiently small $\alpha \neq 0$ we then have $(g + \alpha h)(x_i) \neq 0$ for $i = 1,\dots,k$.

<u>Proof of theorem 4.3:</u> a) follows immediately from b). For the proof of b) we suppose without loss of generality an $f \in U$ has n zeros $x_1<\dots<x_n$ and there are $y_1,\dots,y_n \in M$ with $x_1<y_1<x_2<\dots<x_n<y_n$ and $f(y_i) \neq 0$ for $i = 1,\dots,n$.
Let $I: = \{1,\dots,n\}$, $D: = \{i \in I \mid f(y_{i-1}) \cdot f(y_i) > 0\} \cup \{1\}$, $d: = \# D$.
According to proposition A, let a $g \in U$ with $g(x_i) \neq 0$ for $i \in I$ be fixed and define $R: = \{i \in D \mid g(x_i) \cdot f(y_i) > 0\}$, $r: = \# R$.
For all sufficiently small $\alpha > 0$ and all $i \in I$ we have

$$\operatorname{sign}(f - \alpha g)(y_i) = \operatorname{sign} f(y_i).$$

Let such an $\alpha > 0$ be fixed, and put $h: = f - \alpha g$. For $i \in R$ follows

$$\operatorname{sign} h(x_i) = - \operatorname{sign} g(x_i) = - \operatorname{sign} f(y_i) = - \operatorname{sign} h(y_i),$$

and for $i \in R \setminus \{1\}$ additionally

$$\operatorname{sign} h(y_{i-1}) = \operatorname{sign} h(y_i).$$

We see that if $1 \in R$ holds h has a strong alternation of length 2 in x_1,y_1 and for $i \in R \setminus \{1\}$, h has a strong alternation of length 3 in y_{i-1},x_i,y_i. Besides, h has a strong alternation of length 2 in y_{i-1},y_i for $i \in E: = I \setminus D$. If we now consider increasing indices $i \in I$, it becomes clear that for each $i \in E \cup R$ at least one alternation point of h is added and for $i \in R \setminus \{1\}$ at least two alternation points are added. In the previous arguments $-g$ can be used instead of g. So without loss of generality we may assume $r \ge \frac{d}{2}$. If $r = \frac{d}{2}$ and $1 \in R$, a switch to $-g$ instead of g yields $1 \notin R$. So at least d points may be added to the strong alternation of h when only the indices $i \in R$ are considered. The same holds true for $r \ge \frac{d+1}{2}$. In any case the set $\{x_1,y_1,\dots,x_n,y_n\}$ contains a strong alternation of h of length $n + 1$ since the indices $i \in E$ result in at least $n - d + 1$ more alternation points of h. So we arrive at a contradiction to the hypothesis that U is a weak Haar space.

<u>Corollary 4.4:</u> Under the hypotheses of theorem 4.3, for every $f \in U$ the set Z of zeros of f is the union of m sets $Z_1,\dots,Z_m$ where $m \le n$ and each Z_i is the intersection of M with some interval.
If M is an interval Z consists of at most n intervals.

From theorem 4.3 various conditions may be derived under which a weak Haar space is a Haar space. We begin with a definition:

Definition: A set $M \subset \mathbb{R}$ is said to have property (B) if for every pair $x,y \in M$ with $x < y$ there exists a $z \in M$ with $x < z < y$.

Theorem 4.5: Let $M \subset \mathbb{R}$ be a set with property (B) which does not contain both its infimum and its supremum. Let $U \subset F$ be an n-dimensional weak Haar space and M consist entirely of essential points. If every $f \in U \setminus \{0\}$ has only finitely many zeros, U is a Haar space.

Proof: Let $f \in U \setminus \{0\}$. Between two zeros of f there are infinitely many points of M because of property (B). So all zeros are separated. By theorem 4.3 a) f then has at most n zeros. f cannot have exactly n zeros, say $x_1 < \ldots < x_n$, because f would then vanish left of x_1 and right of x_n (theorem 4.3 b)) and have infinitely many zeros.

Corollary 4.6: Let $M \subset \mathbb{R}$ be a set with property (B) containing neither its infimum nor its supremum. Let $U \subset F$ be an n-dimensional weak Haar space, and assume every $f \in U \setminus \{0\}$ has only finitely many zeros. Then there is a finite subset $E \subset M$ such that $U|_{M \setminus E}$ is a Haar space.

The conditions on M in theorem 4.5 cannot be weakened, as the following examples show:

a) Let $M = [0,\pi]$ or $M = \{-2\pi\} \cup [0,\pi)$, and define $f, g \in F$ by
$f(x) = \sin(x)$, $g(x) = \cos(x)$ for $x \in M$.
$U := \text{span}\,\{f,g\}$ is not a Haar space, but a weak Haar space.

b) Let $M = [0,2\pi]$ or $M = \{-2\pi\} \cup [0,2\pi)$, and U the space of trigonometric polynomials of degree $\le n$ (see also example 5.6).

We shall see in a different context in chapter 10 that examples analogous to a) exist for arbitrary even dimensions $n \ge 2$.
It may have been observed that in example a) every $f \in U$ has a zero while in example b) U contains positive functions. In fact there is a connection between the existence of positive functions and the dimension:

Theorem 4.7: Let $U \subset F$ be an n-dimensional weak Haar space that contains a positive function. If there is an $f \in U$ with n separated zeros, n is odd.

Proof: Let $x_1<\dots<x_n$ be zeros of an $f \in U \setminus \{0\}$, $I := \{1,\dots,n-1\}$ and $y_i \in M \cap (x_i,x_{i+1})$ with $f(y_i) \neq 0$ for $i \in I$. We define

$$P := \{i \in I \mid f(y_i) > 0\},\ p = \# P,$$
$$Q := \{i \in I \mid f(y_i) < 0\},\ q = \# Q.$$

Without loss of generality let $p \geq q$, so $p \geq \frac{n-1}{2}$ because of $p + q = n - 1$. Let $P = \{x_{i_1},\dots,x_{i_p}\}$, $x_{i_1}<\dots<x_{i_p}$, and $g \in U$ be a positive function. For sufficiently small $\alpha > 0$ the function $f - \alpha g \in U$ has a strong alternation of length $2p + 1$ in

$$x_{i_1}, y_{i_1}, x_{i_1+1},\ y_{i_2}, x_{i_2+1},\ \dots, y_{i_p}, x_{i_p+1}.$$

If n were even, from $p \geq \frac{n-1}{2}$ would follow $p \geq \frac{n}{2}$ and $2p + 1 \geq n + 1$, and f would have too long a strong alternation.

The contents of theorems 4.5 and 4.7 may be joined and further refined:

Theorem 4.8: Let M be a set with property (B) and $U \subset F$ an n-dimensional weak Haar space containing a positive function. Let every $f \in U \setminus \{0\}$ have only finitely many zeros. Then U is a Haar space, or all of the following statements hold:

a) M contains its infimum a and its supremum b.
b) n is odd.
c) $U|_{M\setminus\{a\}}$ and $U|_{M\setminus\{b\}}$ are n-dimensional Haar spaces.
d) For all $g \in U$ holds $g(a) \cdot g(b) \geq 0$.
e) $U|_{\{a,b\}}$ has dimension 1.

Proof: a) is an immediate consequence of theorem 4.5.
b) follows from theorem 4.7 since U is no Haar space and the zeros of each $f \in U \setminus \{0\}$ are separated.
c) For every $x \in M$, $U|_{M\setminus\{x\}}$ has dimension n, for otherwise there would exist a $g \in U \setminus \{0\}$ vanishing on $M \setminus \{x\}$ and thus having infinitely many zeros. The statement now follows from theorem 4.5.
d) As U is no Haar space there is an $f \in U \setminus \{0\}$ with n (separated) zeros $x_1<\dots<x_n$. From theorem 4.3 we get $a = x_1$ and $b = x_n$. As f has $n - 1$ zeros on $M \setminus \{a\}$, f changes sign strongly in each zero (chapter 3, exercise 1). Let $y_i \in M \cap (x_i,x_{i+1})$ for $1 = 1,\dots,n-1$ be fixed. So $y_1,\dots,y_{n-1}$ form a strong alternation of f. From b) it is clear that $f(y_1) \cdot f(y_{n-1}) < 0$. If there were a $g \in U$ with $g(a) < 0 < g(b)$ the function $f + \alpha g \in U$ would have a strong alternation of length $n + 1$ in $a,y_1,\dots,y_{n-1},b$ for some $\alpha \neq 0$ of sufficiently small absolute value.

e) If we had $\dim U|_{\{a,b\}} = 2$ there would exist a $g \in U$ with $g(a) \cdot g(b) < 0$.

We remark that a statement analogous to exercise 1 in chapter 3 cannot be expected for weak Haar spaces even if M consists of essential points only. We give two examples:

a) Let $M = \{1,2,3,4,5\}$ and $U = \operatorname{span}\{f_1,f_2,f_3\} \subset F$ with $f_1(2) = f_2(4) = f_3(1) = -f_3(3) = f_3(5) = 1$ and $f_i(j) = 0$ for all other pairs $i,j \in \{1,2,3\} \times M$. $f_1 + f_2 \in U$ has three separated zeros, but no strong change of sign. U is a 3-dimensional weak Haar space.

b) Let $M = \{0,1,2,3,4,5,6\}$ and $U = \operatorname{span}\{f_1,f_2,f_3,f_4\} \subset F$ with $f_1(2) = f_2(4) = f_3(1) = -f_3(3) = f_3(5) = f_3(6) = f_4(1) = 1$ and $f_i(j) = 0$ for all other pairs $i,j \in \{1,2,3,4\} \times M$. U is a 4-dimensional weak Haar space. $f_1 + f_2 \in U$ has three separated zeros in $M \setminus \{\inf M, \sup M\}$, but no strong change of sign.

Exercises: 1) For every n = 1,2,... find a weak Haar space $U \subset C(\mathbb{R})$ of dimension n that contains a function with infinitely many separated zeros. 2) For every n = 1,2,... find an n-dimensional weak Haar space in $C(\mathbb{R})$ that is no Haar space and in which no function has n separated zeros.

5. Periodic Haar spaces

In this chapter we investigate the special case of periodic functions. For $M \subset \mathbb{R}$ a function $f \in F$ is called periodic with period $\tau > 0$ (or shortly: τ-periodic) if for all $x \in M$ we have $x \pm \tau \in M$ and $f(x \pm \tau) = f(\tau)$. We denote by $F_\tau \subset F$ the (sometimes empty) subspace of τ-periodic functions, and put $C_\tau(M) = F_\tau \cap C(M)$.

Definition: An n-dimensional linear space $U \subset F_\tau$ is called a periodic Haar, weak Haar or generalized Haar space on M if for every $a \in M$ the restriction of U to $M \cap [a, a+\tau)$ has the corresponding properties.

Lemma 5.1: Every n-dimensional periodic weak Haar space $U \subset F_\tau$ with at least n + 1 essential points in $M \cap [0,\tau)$ has odd dimension.

Proof: As $U|_{M \cap [0,\tau)}$ is an n-dimensional linear space there exist $x_1,\ldots,x_n \in M$ with $0 \le x_1 < \ldots < x_n < \tau$ such that $U|_{\{x_1,\ldots,x_n\}}$ is a Haar space (lemma 1.1). Let $f \in U$ defined by $f(x_i) = (-1)^i$ for $i = 1,\ldots,n$.

By the hypothesis there is a $p \in M \cap [0,\tau) \setminus \{x_1,\dots,x_n\}$ and a $g \in U$ with $g(p) \neq 0$. For a sufficiently small $\alpha > 0$ the function $h := f + \alpha g$ has a strong alternation of length n in $x_1,\dots,x_n$, and we have $h(p) \neq 0$.
Let $p \in (x_i,x_{i+1})$ for some $i \in \{1,\dots,n-1\}$ or $p \in [0,x_1)$ or $p \in (x_n,\tau)$. Without loss of generality assume $h(p) \cdot h(x_{i+1}) > 0$ or $h(p) \cdot h(x_1) > 0$ otherwise. Now consider the points
$x_{i+1},\dots,x_n,x_1 + \tau,\dots,x_i + \tau,p + \tau \in [x_{i+1},x_{i+1} + \tau)$ or
$x_1,\dots,x_n,p + \tau \in [x_1,x_1 + \tau)$ or $x_1,\dots,x_n,p \in [x_1,x_1 + \tau)$. If we had $h(x_1) \cdot h(x_n) < 0$, in each of these cases h would have a strong alternation of length n + 1 in the points considered, contradicting the hypothesis. The statement now follows from $h(x_1) \cdot h(x_n) > 0$ and $(-1)^{n-1} h(x_1) \cdot h(x_n) > 0$, where the last inequality is due to the fact that h has a strong alternation in $x_1,\dots,x_n$.

Corollary 5.2: Every n-dimensional periodic Haar space $U \subset F_\tau$ with $\# (M \cap [0,\tau)) \geq n + 1$ has odd dimension.

Lemma 5.3: Let $U \subset F_\tau$ be an n-dimensional periodic weak Haar space, and assume M consists of essential points only. Then for every $f \in U$ there is an $a \in \mathbb{R}$ such that f has at most n - 1 separated zeros on $M \cap [a,a+\tau)$.

Proof: By theorem 4.3 f has at most n separated zeros in $M \cap [0,\tau)$. If f has exactly n separated zeros $x_1<\dots<x_n$ in $[0,\tau)$, in addition f vanishes on $M \cap ([0,x_1] \cup [x_n,\tau))$. Let $a \in M \cap (x_1,x_2)$ with $f(a) \neq 0$. Then f has only n - 1 separated zeros on $M \cap [a,a + \tau)$.

Theorem 4.5 directly implies:

Theorem 5.4: If the hypotheses of lemma 5.3 hold, M has property (B) and every $f \in U \setminus \{0\}$ has only finitely many zeros, then U is a periodic Haar space.

Theorem 4.8 may be reformulated:

Theorem 5.5: Under the hypotheses of theorem 4.8 either U is a Haar space, or the following statements hold:

a) M contains its infimum a and its supremum b.
b) With $\tau = b - a$ the restrictions $U|_{M\setminus\{b\}}$ and $U|_{M\setminus\{a\}}$ can be extended to τ-periodic n-dimensional Haar spaces $\tilde{U}$ on $\tilde{M}$, if $\tilde{M}$ is defined by $\tilde{M} = \{x + k\tau | k \in \mathbb{Z} \text{ and } x \in M\}$, and $\tilde{f} \in \tilde{U}$ by $\tilde{f}(x) = \tilde{f}(y)$ for $x,y \in M$ with $x = y (\mathrm{mod}\ \tau)$ and $\tilde{f}|_M = f$ for some $f \in U$.

Proof: $\tilde{U}$ obviously is a τ-periodic generalized Haar space. Now suppose an $\tilde{f} \in \tilde{U}$ has a strong alternation of length n+1 on $\tilde{M} \cap [p,p+\tau)$ for some $p \in \tilde{M}$, say $x_1<\ldots<x_{n+1}$. Without loss of generality let $p \in M \setminus \{b\} = \tilde{M} \cap [a,a+\tau)$. If $x_{n+1} < a+\tau$ we immediately get a contradiction. If $a+\tau < x_1$, the points $x_1-\tau,\ldots,x_{n+1}-\tau \in M \setminus \{b\}$ form a strong alternation of f of length n+1. If $x_i < a+\tau < x_{i+1}$ for some i, the points $x_{i+1}-\tau,\ldots,x_{n+1}-\tau, x_1,\ldots,x_i \in M \setminus \{b\}$ form a strong alternation of f of length n+1, since n is odd and we have $f(x_1)\cdot f(x_{n+1}) < 0$.

The simplest example of a periodic Haar space is formed by the trigonometric polynomials:

Example 5.6: Let $M = \mathbb{R}$, $n \geq 1$, and for $v = 0,1,\ldots,n$ let $f_v, g_v \in C(\mathbb{R})$ be defined by

$$f_v(x) = \cos(vx)$$
$$g_v(x) = \sin(vx), x \in \mathbb{R}.$$

$U := \operatorname{span}\{f_v, g_v\}_{v=0}^{n}$ is an (2n+1)-dimensional linear space of 2π-periodic functions.

Let $f = \sum_{v=0}^{n} \alpha_v f_v + \beta_v g_v$. The substitution $\cos(vx) + i\sin(vx) = e^{ivx}$ leads to

$$f(x) = \frac{1}{2}\sum_{v=0}^{n} (\alpha_v - i\beta_v)\, e^{ivx} + (\alpha_v + i\beta_v)\, e^{-ivx}.$$

If we put $z := e^{ix}$, we get

$$f(x) = z^{-n} P(z),$$

where P is a complex polynomial of degree $\leq 2n$. As every $P \neq 0$ has at most 2n zeros, f has at most 2n zeros on $[a,a+2\pi)$ for every $a \in \mathbb{R}$. By theorem 3.5 U is a Haar space on every interval $[a,a+2\pi)$.

Exercises: 1) Show that the definition of periodic weak Haar or Haar spaces at the beginning of this chapter is consistent insofar as for every basis of U, the determinant corresponding to $(x_1,\ldots,x_n) \in \Delta_n(M)$ has weakly or strongly constant sign for all $x_1,\ldots,x_n \in M \cap [a,a+\tau)$ and all $a \in M$.

2) A bijective mapping $h: [0,\tau) \to S^1$ is defined by $h(x) = e^{i\frac{2\pi}{\tau}x}$ for $x \in [0,\tau)$, if here we consider S^1 as a subset of the complex plane. The family T of subsets of $[0,\tau)$ defined by

$$T = \{A \subset [0,\tau) \mid A = (a,b) \text{ or } A = [0,a) \cup (b,\tau) \text{ for } a,b \in (0,\tau)\}$$

generates a topology on $[0,\tau)$, the "periodic" topology. Show that h is a homeomorphism with respect to this topology.

(Thus a Haar space $U \subset C(S^1)$ may be considered as a periodic Haar space in C_τ, especially in $C_{2\pi}$).

6. Sign changes, prescribed zeros

In this chapter we shall investigate the possibility of constructing functions in Haar spaces with simple or double zeros (terms still to be defined) in prescribed points.

Though it may not be absolutely necessary for the derivation of our main result, it is perhaps helpful first to discuss the various types of sign changes encountered in this situation.

Definition: Let $f \in F$ and $k \in \{0,1,2,\ldots\}$. f is said to have (exactly) k sign changes without zeros iff there exist (exactly) k pairs $x_i, y_i \in M$, $i=1,\ldots,k$, with $x_1 < y_1 \leq x_2 < y_2 \leq \ldots \leq x_k < y_k$,

$f(x_i) \cdot f(y_i) < 0$ for $i = 1,\ldots,k$ and such that

$f|_{M \cap [x_i, y_i]}$ has no zero and no weak (strong) alternation of length 3.

If M is an interval and $f \in C(M)$, f clearly has no sign changes without zeros.

As an extension of lemma 3.1 we have

Lemma 6.1: Let $U \subset F$ be an n-dimensional linear space. Then the following properties are equivalent:

a) U is a Haar space.

b) No $f \in U \setminus \{0\}$ has $n - k$ zeros and k sign changes without zeros for any $k \in \{0,1,\ldots,n\}$.

Proof: b) => a): U obviously is a generalized Haar space. Suppose an $f \in U$ has a strong alternation of length $n + 1$ in $(t_1,\ldots,t_{n+1}) \in \Delta_{n+1}(M)$. Then $f|_{M \cap [t_i, t_{i+1}]}$ has a zero or at least one sign change without a zero for $i=1,\ldots,n$. So f has at least $n - k$ zeros and k sign changes without zeros for some $k \in \{0,1,\ldots,n\}$ in contradiction to the hypothesis.

a) => b): Suppose an $f \in U \setminus \{0\}$ has $n - k$ zeros $z_1 < \ldots < z_{n-k}$ and k sign changes without zeros for some $k \in \{0,1,\ldots,n\}$. Excluding the trivial case $k = 0$, let $x_i, y_i \in M, i=1,\ldots,k$, be such that $x_1 < y_1 \leq x_2 < y_2 \leq \ldots \leq x_k < y_k, f(x_i) \cdot f(y_i) < 0$ and $f|_{M \cap [x_i, y_i]}$ has no zero and no alternation of length 3 for $i=1,\ldots k$. One easily checks that the set

$\{z_1,\dots,z_{n-k},x_1,y_1,\dots,x_k,y_k\}$ contains at least n+1 points forming a weak alternation of f in contradiction to a).

For $f \in F$, let the sets SZ(f) and DZ(f) of "simple" and "double" zeros be defined by

$Z(f) = \{x \in M \mid f(x)=0\}$,

$DZ(f) = \{x \in Z(f) \mid$ there exist $u,v \in M$ with $u<x<v$ such that f has constant nonzero sign on $(M \cap [u,v]) \setminus \{x\}\}$,

$SZ(f) = Z(f) \setminus DZ(f)$.

Denote the cardinalities of SZ(f) and DZ(f) by $\sigma(f)$ and $\delta(f)$, respectively.

<u>Lemma 6.2:</u> Let $U \subset F$ be an n-dimensional Haar space. Then $\sigma(f) + 2\delta(f) \leq n - 1$ holds for every $f \in U \setminus \{0\}$.

<u>Proof:</u> Suppose there are an $f \in U \setminus \{0\}$ and $(s_1,\dots,s_k) \in \Delta_k(M)$, $(d_1,\dots,d_l) \in \Delta_l(M)$ with $SZ(f) = \{s_1,\dots,s_k\}$, $DZ(f) = \{d_1,\dots,d_l\}$ and $k + 2l \geq n$.

Let indices $i_0,\dots,i_m$ and $j_1,\dots,j_m$ be given by

$$s_1 < \dots < s_{i_0} < d_1 < \dots < d_{j_1} < s_{i_0+1} < \dots < s_{i_1} < d_{j_1+1} < \dots$$
$$\dots < s_{i_m} < d_{j_m+1} < \dots < d_l < s_{i_m+1} < \dots < s_k.$$

For $\nu=1,\dots,l$, let $u_\nu,v_\nu \in M$ be such that $u_\nu < d_\nu < v_\nu$ and f has constant nonzero sign on $(M \cap [u_\nu,v_\nu]) \setminus \{d_\nu\}$. One checks that the set $\{s_1,\dots,s_{i_0},u_1,d_1,v_1,\dots,u_{j_1},d_{j_1},v_{j_1},s_{i_0+1},\dots,s_{i_1},u_{j_1+1},d_{j_1+1},v_{j_1+1},\dots$ contains at least $k + 2l + 1 \geq n + 1$ points forming a weak alternation of f, a contradiction.

For $f \in F$ denote by $\tau(f)$ the number of sign changes of f without zeros. Then lemma 6.1 can be sharpened to:

<u>Lemma 6.3:</u> Let $U \subset F$ be an n-dimensional linear space. Then the following properties are equivalent:

a) U is a Haar space.

b) $\sigma(f) + 2\delta(f) + \tau(f) \leq n - 1$ holds for all $f \in U \setminus \{0\}$.

<u>Proof:</u> If b) holds, a) is implied by lemma 6.1.

If a) holds, but b) does not for some $f \in U \setminus \{0\}$, a weak alternation of f of length $\sigma(f) + 2\delta(f) + \tau(f) + 1$ can be found in a fashion completely analogous to the proof of lemma 6.2.

For the proof of the main result of this section we may use the following corollary of lemma 6.3 which can also be proved directly:

Lemma 6.4: Let $U \subset F$ be an n-dimensional Haar space, and $f \in U$ with $\sigma(f) + 2\delta(f) = n - 1$. Then f has no sign changes without zeros.

Theorem 6.5: Let $U \subset F$ be an n-dimensional Haar space and $s_1,\ldots,s_k,d_1,\ldots,d_l \in M$ pairwise distinct points with $s_1 < \ldots < s_k$, $a < d_1 < \ldots < d_l < b$ and $0 \le k + 2l \le n - 1$, where $a: = \inf M$, $b: = \sup M$. Then there is an $f \in U$ with $SZ(f) = \{s_1,\ldots,s_k\}$, $DZ(f) = \{d_1,\ldots,d_l\}$ and no sign changes without zeros if at least one of the following conditions holds:

a) $n - (k+2l)$ is odd;
b) $a,b \notin M$;
c) $a,b \in M$ and $a < s_1, s_k < b$;
d) $a,b \in M$, $a = s_1, s_k = b$ and $M \setminus \{a,b\}$ contains no smallest and no largest element.
e) $a \in M, b \notin M, a = s_1$ and $M \setminus \{a\}$ contains no smallest element.

So a function $f \in U$ with exactly k simple and l double zeros, $0 \le k + 2l \le n - 1$, in prescribed points and with no sign changes without zeros, exists if U contains an (n-1)-dimensional Haar space. Another special case is that U contains a positive function if n is odd or if $a,b \in M$ or if $a,b \notin M$.
As some confusion exists in the literature to this point, we wish to emphasize that the statement of theorem 6.5 does not hold in general, but may be violated in the following cases:

1) $a,b \in M, n - (k+2l)$ even, and $\begin{cases} a = s_1, s_k < b \text{ or } a < s_1, s_k = b \text{ or} \\ a < \min(M \setminus \{a\}) \text{ or } \max(M \setminus \{b\}) < b \end{cases}$

2) $a \in M, b \notin M, n - (k+2l)$ even, and $\begin{cases} a < s_1 \text{ or} \\ a < \max(M \setminus \{a\}) \end{cases}$

For reasons of presentation, corresponding examples are deferred to chapter 10.

Proof of theorem 6.5: a) It is sufficient to find $\hat{g},\hat{h} \in U$ with no sign changes without zeros, and

$$\hat{g}(x) \cdot \hat{h}(x) \geq 0 \quad \text{for all } x \in M,$$
$$SZ(\hat{g}) = SZ(\hat{h}) = \{s_1,\dots,s_k\},$$
$$DZ(\hat{g}) \cap DZ(\hat{h}) = \{d_1,\dots,d_l\},$$
$$\sigma(\hat{g}) + 2\delta(\hat{g}) = \sigma(\hat{h}) + 2\delta(\hat{h}) = n-1,$$

because $f := \hat{g} + \hat{h}$ then has the desired properties. So we need to consider only the case $k + 2l = n - 1$.

For each $i \in \{1,\dots,l\}$, let $\{u_i^\nu\}_{\nu=1}^\infty$ be an increasing and $\{v_i^\nu\}_{\nu=1}^\infty$ a decreasing sequence in M with

$u_i^\nu \to u_i := \sup\{M\cap(-\infty,d_i)\}, v_i^\nu \to v_i := \inf\{M\cap(d_i,\infty)\}$ for $\nu \to \infty$.

If $\|\ \|$ denotes some norm on U (all norms on U are equivalent), let $\{g_\nu\}_{\nu=1}^\infty$ and $\{h_\nu\}_{\nu=1}^\infty$ be sequences in $S := \{f\in U \mid \|f\|=1\}$ such that g_ν vanishes on $G_\nu := \{s_1,\dots,s_k,u_1^\nu,d_1,\dots,u_l^\nu,d_l\}$ and h_ν on $H_\nu := \{s_1,\dots,s_k,d_1,v_1^\nu,\dots,d_l,v_l^\nu\}$, $\nu = 1,2,\dots$. Without loss of generality we may assume that each G_ν and H_ν consists of $n-1$ distinct points and

$$G_\nu \cap (u_i^\nu,d_i) = \emptyset = H_\nu \cap (d_i,v_i^\nu)$$

for $i=1,\dots,l$ and all ν.

As S is compact, $\{g_\nu\}$ and $\{h_\nu\}$ have convergent subsequences; for simplicity let us assume

$$\|g_\nu - g\| \to 0, \ \|h_\nu - h\| \to 0 \text{ for } \nu \to \infty \text{ and suitable } g,h \in S.$$

By chapter 3, exercise 1, every g_ν and every h_ν change sign exactly in their zeros ($= G_\nu$ or $= H_\nu$ respectively). As norm convergence implies pointwise convergence, we get

$$DZ(g) = \{d_i \mid u_i = d_i\}, \ DZ(h) = \{d_i \mid d_i = v_i\},$$
$$SZ(g) = \{s_1,\dots,s_k\} \cup \bigcup_{u_i<d_i} \{u_i,d_i\}, SZ(h) = \{s_1,\dots,s_k\} \cup \bigcup_{d_i<v_i} \{d_i,v_i\},$$

and g and h have no sign changes without zeros (this last statement may be deduced directly from the construction or via lemma 6.4). So g and h or g and -h have weakly the same sign throughout M, say $g(x) \cdot h(x) \geq 0$ for $x \in M$. For $f := g + h$ we get

$\{s_1,\dots,s_k\} \subset SZ(f)$ and $Z(f) \subset \{s_1,\dots,s_k,u_1,d_1,v_1,\dots,u_l,d_l,v_l\}$.

No u_i with $u_i < d_i$ can be a zero of f since we have $h(u_i) \neq 0$; analogously no v_i with $d_i < v_i$ lies in $Z(f)$, so $Z(f) \subset \{s_1,\dots,s_k,d_1,\dots,d_l\} =: A$. Now let $i \in \{1,\dots,l\}$ be fixed and $l := \max\{A\cap(-\infty,d_i)\}, r := \min\{A\cap(d_i,\infty)\}$. By our construction the sets $M \cap (l,d_i)$ and $M \cap (d_i,r)$ are not empty, g has constant non-zero sign on $(M \cap (l,r)) \setminus \{u_i,d_i\}$, and h has the same sign on $(M \cap (l,r)) \setminus \{d_i,v_i\}$. So f has a double zero in d_i regardless whether $u_i < d_i$, or $u_i = d_i$ and $d_i = v_i$, or $d_i < v_i$. As f has no sign changes without zeros, f is a function in U with the desired properties.

b) As U contains an (n-1)-dimensional Haar space by theorem 7.7 - which does not depend on any results of this chapter -, the assertion follows immediately from a).

c) We may assume $n - (k + 2l)$ to be even. By a) there exist $f_a, f_b \in U$ with $DZ(f_a) = DZ(f_b) = \{d_1,\dots,d_l\}$, $SZ(f_a) = \{a,s_1,\dots,s_k\}$, $SZ(f_b) = \{s_1,\dots,s_k,b\}$ and no sign changes without zeros. Then $f := f_a + f_b$ or $f := f_a - f_b$ has the desired properties.

d) The restriction to $M \setminus \{a,b\}$ of the space

$$\overline{X} := \{f \in U | f(a) = f(b) = 0\}$$

is an (n-2)-dimensional Haar space. Now apply b).

e) The restriction to $M \setminus \{a\}$ of the space

$$\overline{X} := \{f \in U | f(a) = 0\}$$

is an (n-1)-dimensional Haar space. Now apply b).

Corollary 6.6: Let the hypotheses of theorem 6.5 be fulfilled. Then there exists an $f \in U$ such that f has no sign changes without zeros, $DZ(f) = \{d_1,\dots,d_l\}$ and $\{s_1,\dots,s_k\} \subset SZ(f) \subset \{a,s_1,\dots,s_k\}$ or $\{s_1,\dots,s_k\} \subset SZ(f) \subset \{s_1,\dots,s_k,b\}$.

Proof: If one of the conditions a) to e) in theorem 6.5 holds, nothing has to be proved. Otherwise, let us first assume $a,b \in M$. For $a = s_1, s_k < b$ theorem 6.5 a) gives an f with $SZ(f) = \{s_1,\dots,s_k,b\}$, for $a < s_1, s_k = b$ theorem 6.5 a) gives an f with $SZ(f) = \{a,s_1,\dots,s_k\}$. For $a = s_1, s_k = b$ and $a < \min(M \setminus \{a\})$ consider the restriction of $V := \{f \in U | f(a) = 0\}$ to $M \setminus \{a\}$ and use induction over n. Now assume $a \in M$, $b \notin M$ without loss of generality. For $a < s_1$ theorem 6.5 a) gives an f with $SZ(f) = \{a,s_1,\dots,s_k\}$. For $a = s_1 < \min(M \setminus \{a\})$ again consider $V|_{M \setminus \{a\}}$ and use induction over n.

For $k \in \mathbb{N}$ we define $\overline{\Delta}_k(M) := \{(t_1,\dots,t_k) \in M^k | t_1 \leq \dots \leq t_k\}$.

Theorem 6.7: Let $U \subset F$ be an n-dimensional Haar space, $\theta_\mu = (z_1^\mu,\dots,z_k^\mu) \in \overline{\Delta}_k(M)$ for $\mu = 1,2,\dots$ with

$\theta_\mu \to \theta := (z_1,\ldots,z_k) \in \bar{\Delta}_k(\bar{M})$ for $\mu\to\infty$, $z_0^\mu := z_0 := -\infty$ and $z_{k+1}^\mu := z_{k+1} := \infty$ for $\mu = 1,2,\ldots$. Then there is an $f \in U\setminus\{0\}$ and a subsequence $\left\{\theta_{\mu_j}\right\}_{j=1}^\infty$ such that

$$\lim_{j\to\infty} f(z_i^{\mu_j}) = 0 \text{ for } i = 1,\ldots,k$$

and

$$(-1)^i f(x) \geq 0 \quad \text{for } x \in (z_{i-1},z_i) \cap M,\ i = 1,\ldots,k+1.$$

Proof: By corollary 6.6, for every μ there is an $f_\mu \in U\setminus\{0\}$ with $(z_1^\mu,\ldots,z_k^\mu) \subset Z(f_\mu)$ and $(-1)^i f_\mu(x) \geq 0$ for $x \in (z_{i-1}^\mu,z_i^\mu) \cap M$, $i = 1,\ldots,n$. If $\|\ \|$ is some fixed norm on U, we may without loss of generality assume $f_\mu \in S := \{h\in U \mid \|h\| = 1\}$ for all μ. As U has finite dimension, S is compact with respect to the topology generated by the norm, and $\{f_\mu\}$ contains a convergent subsequence $\{f_{\mu_j}\}_{j=1}^\infty$, say $f_{\mu_j} \to f \in S$ for $j \to \infty$. One easily checks that f has the desired properties.

Exercises: 1) Let $M \subset \mathbb{R}$ and $U \subset F$ be an n-dimensional linear space, $z_0 := -\infty$, $z_n := \infty$. Show the following converse of exercise 1 in chapter 3:
If for every $(z_1,\ldots,z_{n-1}) \in \Delta_{n-1}(M)$ there is an $f \in U \setminus \{0\}$ with $Z(f) = \{z_1,\ldots,z_{n-1}\}$ and $(-1)^i f(x) > 0$ for $x \in (z_{i-1},z_i) \cap M$, $i = 1,\ldots,n$, then U is a Haar space. (GOPINATH and KURSHAN 1977a, theorem 1).
Hint: Show first U is a generalized Haar space. For the rest use lemma 6.1.
2) Give a detailed proof of lemma 3, a) => b).
3) *Instead of invoking theorem 7.7, prove theorem 6.5 b) directly.
Hint: Consider the case n - (k+2l) even and use the methods developed in the proof of part a). Take a sequence $\{f_\nu\}_1^\infty$ in U with each f_ν having n - 1 zeros (counting double zeros twice) and the "superfluous" zeros moving to a or to b.

III. Subspaces

In this section we shall mainly deal with the question under which conditions a Haar, weak Haar or generalized Haar space contains subspaces of the same kind.

7. Existence theorems

In a linear space $U \subset F$ let Z_k denote the set of all $f \in U \setminus \{0\}$ that have at least k zeros, $k = 0,1,2,\ldots$.

Theorem 7.1: Let $U \subset F$ be an n-dimensional generalized Haar space, and $k \in \{0,1,\ldots,n\}$ fixed. Then the following statements are equivalent:
a) U contains a k-dimensional generalized Haar space.
b) U^* contains an (n-k)-dimensional generalized Haar space W with the property that for every $f \in Z_k$ there is a $t^* \in W$ with $t^*(f) \neq 0$.

Proof: a) => b): Let $V \subset U$ be a k-dimensional generalized Haar space. Let $W := V^\perp$ be the (n-k)-dimensional subspace of U^* that annihilates U. As Z_k and V have no common element, no $f \in Z_k$ can be annihilated by all linear functionals in W.
b) => a): Let $V := W^\perp = \{f \in U \mid t^*(f) = 0 \text{ for all } t^* \in W\}$. We have $\dim V = k$. No $f \in V \setminus \{0\}$ has more than $k - 1$ zeros, for otherwise we would have $f \in Z_k$, and a $t^* \in W$ would exist with $t^*(f) \neq 0$.

For the following theorem we introduce the concept of extendability: The domain of definition of U, M, can be extended by a point x if there is an n-dimensional space $\hat{U}$ of the same kind, defined on $M \cup \{x\}$, whose restriction to M is just U.

Corollary 7.2: Let $U \subset F$ be an n-dimensional generalized Haar space, $n \geq 1$. Then the following properties are equivalent:
a) U contains an (n-1)-dimensional generalized Haar space.
b) There is a $t^* \in U^*$ with $t^*(f) \neq 0$ for all $f \in Z_{n-1}$.
c) M may be extended by one point.
d) Let $f_1,\ldots, f_n$ be a basis of U and $\phi(t) := (f_1(t),\ldots,f_n(t))$ for $t \in M$. There is an $x \in \mathbb{R}^n$ such that every (n-1)-dimensional hyperplane through x and 0 intersects $\phi(M)$ in at most $n - 2$ points.

Proof: The equivalence of a) and b) follows from theorem 7.1.
c) => a): Let $\hat{U}$ be an extension of U to a set $M \cup \{p\}$, $p \notin M$. Now define $\hat{V} := \{f \in \hat{U} \mid f(p) = 0\}$. The restriction of $\hat{V}$ to M is an (n-1)-dimensional generalized Haar space (see lemma 1.3).

b) => c): Let $p \notin M$, and for all $f \in U$ define $f(p) := t^*(f)$. If an $h \in U \setminus \{0\}$ had n zeros on $M \cup \{p\}$, we would have $h(p) = 0$ and $h \in Z_{n-1}$ in contradiction to b).
c) <=> d): We first state a lemma which follows directly from lemma 1.2 d):

Lemma 7.3: Let $f_1, \ldots, f_n$ be a basis of a linear space $U \subset F$, and $\phi = (f_1, \ldots, f_n)$. Then the following statements are equivalent:
a) U is a generalized Haar space.
b) For every set of n pairwise different points $x_1, \ldots, x_n \in M$, $\phi(x_1), \ldots, \phi(x_n)$ are linearly independent.
c) No (n-1)-dimensional hyperplane $H_a = \{z \in \mathbb{R}^n \mid \langle z,a \rangle = 0\}$, $a \in \mathbb{R}^n$, intersects $\phi(M)$ in n points.

If now 7.2 c) is fulfilled, say, $\hat{U}$ an extension of U to $M \cup \{p\}$, $p \notin M$, put $x := \phi(p)$ and apply lemma 7.3.
If 7.2 d) holds, for a $p \notin M$ and all $f \in U$ define $f(p)$ by $\phi(p) := x$. For $\hat{M} := M \cup \{p\}$ the statement of lemma 7.3 c) is true, and 7.2 c) follows immediately.

Theorem 7.4: Let M be a finite set and $U \subset F$ an n-dimensional generalized Haar space, $n \geq 1$. Then U contains an (n-1)-dimensional generalized Haar space.

Proof: The set of hyperplanes through 0 that intersect $\phi(M)$ in n-1 points is finite and cannot occupy all of $\mathbb{R}^n$. Now apply Corollary 7.2 d).

We now focus our attention on totally ordered domains M, that is, weak Haar spaces and Haar spaces. Again we assume $M \subset \mathbb{R}$ with little, if any loss of generality (see chapter 3).
If two norms $\|\ \|_1$ and $\|\ \|_2$ are defined on a finite-dimensional linear space U there exist constants $\gamma, \delta \in \mathbb{R}$ such that

$$\|x\|_1 \leq \gamma \|x\|_2 \text{ and } \|x\|_2 \leq \delta \|x\|_1$$

for all $x \in U$. This implies that all norms on U generate the same topology. In the following let the linear spaces under consideration be endowed with this topology.

Lemma 7.5: Let U be an n-dimensional linear space and $\{B_k\}_{k=1}^{\infty}$ an antitone sequence of closed subsets with $B_i \subset B_j$ for $i > j$. Assume that each B_k contains a linear space U_k of dimension $m \leq n$. Then $B := \bigcap_{k=1}^{\infty} B_k$ contains an m-dimensional linear space, too.

Proof: Let U be endowed with an inner product and the corresponding norm $\| \ \|$. For each $k = 1,2,\ldots$ let $e_k^1,\ldots,e_k^m$ be an orthonormal basis of U_k, $E_k := (e_k^1,\ldots,e_k^m) \in U^m$ and $N := \bigcup_{k=1}^{\infty} E_k$.
We have $E_i \in B_j \times\ldots\times B_j$ for $i \geq j$. As

$S := \{(x,\ldots,x^m) \in U^m \mid \|x^1\| = \ldots = \|x^k\| = 1\}$ is compact N has a cluster point $E := \{e^1,\ldots,e^m\} \in S$. Since $B_j \times\ldots\times B_j$ is closed for every j, we get $E \in B_j \times\ldots\times B_j$ for all j and so $E \in B \times\ldots\times B$, i.e., $e^1,\ldots,e^m \in B$. As the inner product is continuous, $e^1,\ldots,e^m$ are orthonormal. From the continuity of the norm and $U_k \subset B_k$ for all k, it is clear that the linear space spanned by $e^1,\ldots,e^m$ lies in B.

Theorem 7.6: Let $U \subset F$ be an n-dimensional weak Haar space, $n \geq 1$. Then U contains an $(n-1)$-dimensional weak Haar space.

Proof: Let $T \subset M$ be the subset of the essential points with respect to U. We denote by A the set of all $f \in U$ with a strong alternation of length n on M or, more precisely, on $M \cap T$.
For all $x \in T$ the spaces

$U_x := \{f \in U \mid f(x) = 0\}$

are $(n-1)$-dimensional linear spaces. Furthermore we define for $x \in T$:

$A_x := \{f \in U \mid f$ has a strong alternation of length n in $M \cap [x,\infty)\}$,

$B_x := U \setminus A_x$.

We suppose for some $x \in T$ the intersection $A_x \cap U_x$ is not empty, say $h \in A_x \cap U_x$. Let $x_1 < \ldots < x_n$ be a strong alternation of $h|_{M\cap[x,\infty)}$ of length n, and without loss of generality assume $h(x_1) > 0$. Let $f \in U$ be a function with $f(x) = 1$. For sufficiently small $\varepsilon > 0$, $h - \varepsilon f$ has a strong alternation of length $n+1$ in $x, x_1,\ldots, x_n$, a contradiction. So we get $U_x \subset B_x$ for all $x \in T$.
If T contains its infimum a, we have $A_a = A$. Then U_a is an $(n-1)$-dimensional weak Haar space. So for the following we assume $\inf T \notin T$.
We note that every set A_x is open, so every B_x is closed. To show this we take an $f \in A_x$ with a strong alternation $x_1 < \ldots < x_n$ on $M \cap [x,\infty)$. Let $\alpha := \min\{|f(x_i)| \mid i = 1,\ldots,n\}$, so $\alpha > 0$. We augment the set $\{x_1,\ldots,x_n\}$ by at most $n-1$ points $p_1,\ldots, p_k \in M$ such that for $N := \{x_1,\ldots,x_n,p_1,\ldots,p_k\}$ the restriction $U|_N$ has dimension n. Thus the maximum norm on $U|_N$ is a norm on U. If $g \in U$ now is any function whose maximum norm on $U|_N$ is smaller than α, $f + g$ lies in A_x. This means that an open neighbourhood of f is contained in A_x.
Obviously we have $B_y \subset B_x$ for $y < x$.
Let $\{x_\nu\}_1^\infty$ be a monotonously decreasing sequence in T with $\lim_{\nu\to\infty} x_\nu = \inf T$.

We see that for $\{B_{x_\nu}\}_1^\infty$ and $\{U_{x_\nu}\}_1^\infty$ the hypotheses of lemma 7.5 are fulfilled. So

$B := \bigcap_{\nu=1}^{\infty} B_{x_\nu} = U \setminus \bigcup_{\nu=1}^{\infty} A_{x_\nu}$ contains an (n-1)-dimensional linear space V.

The set $\bigcup_{\nu=1}^{\infty} A_{x_\nu}$ consists of all $f \in U$ that have a strong alternation of length n on $M \cap (\inf T, \infty)$. As $\inf T \in T$ was assumed, we get $A = \bigcup_{\nu=1}^{\infty} A_{x_\nu}$. So $V \cap A$ is empty, and V is a weak Haar space.

<u>Theorem 7.7:</u> Let $M \subset \mathbb{R}$ be a set which contains neither its infimum nor its supremum, and $U \subset F$ an n-dimensional Haar space, $n \geq 1$. Then U contains an (n-1)-dimensional Haar space.

<u>Proof:</u> By theorem 7.6 U contains an (n-1)-dimensional weak Haar space V. We first suppose M has an unessential point x with respect to V, that is, $f(x) = 0$ for all $f \in V$. Now let $x_1, \ldots, x_n \in M$ be points with $x_1 < \ldots < x_{n-1} < x < x_n$. By lemma 4.1 c) there exists a $g \in V \setminus \{0\}$ with $g(x_1) = \ldots = g(x_{n-2}) = 0$ and weakly constant sign on $M \cap (x_{n-2}, \infty)$. But then g has a weak alternation of length n + 1 in $x_1, \ldots, x_{n-1}, x, x_n$ in contradiction to the hypothesis.
So all points of M are essential with respect to V. If an $h \in V \setminus \{0\}$ had n - 1 zeros $x_1 < \ldots < x_{n-1}$, h would vanish left of x_1 and right of x_{n-1} because of theorem 4.3 b), and h would have infinitely many zeros. Thus V is a Haar space.

Because of the importance of theorem 7.7 we add another proof which does not recur to theorem 4.3:

<u>Second proof:</u> By theorem 7.6 U contains an (n-1)-dimensional weak Haar space V. We suppose an $f \in V \setminus \{0\}$ has n - 1 zeros $x_1 < \ldots < x_{n-1}$. Let $x_0, x_n \in M$ with $x_0 < x_1$, $x_{n-1} < x_n$. From $\dim U|_{\{x_0,\ldots,x_{n-1}\}} = n$ follows $\dim V|_{\{x_0,\ldots,x_{n-1}\}} = n - 1$. So there is an $i \in \{1,\ldots,n-1\}$ such that $V|_{\{x_0,\ldots,x_{i-1},x_{i+1},\ldots,x_{n-1}\}}$ is an (n-1)-dimensional Haar space.
Without loss of generality assume $f(x_0) > 0$. As U is a Haar space exercise 1 in chapter 3 yields $(-1)^n f(x_n) < 0$. Let $g \in V$ be defined by

$$g(x_j) = \begin{cases} (-1)^j & \text{for } j = 0, 1, \ldots, i - 1 \\ (-1)^{j+1} & \text{for } j = i + 1, \ldots, n - 1. \end{cases}$$

For sufficiently small $\varepsilon > 0$, $f + \varepsilon g \in V$ has a strong alternation of length n in $x_0, \ldots, x_{i-1}, x_{i+1}, \ldots, x_n$ in contradiction to the fact that V is a weak Haar space.

If M contains one or both of its endpoints an n-dimensional Haar space need no longer have an (n-1)-dimensional Haar subspace. Such examples will be given in chapter 10.
It does not seem to be known if the statement of theorem 7.7 holds for finite M, too. We guess a fairly simple argument might exist in this case.

Exercises: 1) One step in the proof of theorem 7.6 was to show that A_x is open. Why is it impossible to use the maximum norm? Give suitable counterexamples.
2) *If M has property (B), theorem 7.7 can be derived without using of theorem 7.6 or other properties of weak Haar spaces. Give the details.

8. Characterizations of Markov systems

Beyond the results of the last chapter, the question naturally arises under what conditions an n-dimensional weak Haar space U not only contains an (n-1)-dimensional space of the same kind, but spaces U_i with $\dim U_i = i$, $i = 1,\ldots, n-1$, and $U_i \subset U_2 \subset \ldots \subset U_n$.
For reasons that will become apparent later, corresponding bases are given a special name:

Definition: a) A (weak; generalized) Čebyšev system $f_1,\ldots, f_n \in F$ is called (weak; generalized) Markov system iff $f_1,\ldots, f_i$ form a (weak; generalized) Čebyšev system for $i = 1,\ldots, n$.
b) A (weak;generalized) Markov system $f_1,\ldots, f_n \in F$ is called normed iff $f_1 \equiv 1$.

The formulation of theorems 7.4, 7.6, 7.7 in terms of Markov systems is left to the reader.
Before proceeding we introduce a new type of spaces, so-called oscillation spaces. In the frame of this book these spaces do not constitute an independent concept though they might be interesting enough to stimulate further investigations. Here they mainly serve as a means to obtain simple formulations.

Definition: Let $f \in F$. Points $x_1,\ldots, x_k \in M$ are called a strong (weak) oscillation of f of length k, iff
a) $x_1 <\ldots< x_k$ and
b) $(-1)^i\,[f(x_{i+1})-f(x_i)]$ is positive (nonnegative) for $i = 1,\ldots, k-1$ or negative (nonpositive) for $i = 1,\ldots, k-1$.

Obviously we have

Lemma 8.1: A strong (weak) alternation of an $f \in F$ of length k is a strong (weak) oscillation of f of length k.

Definition: Let $U \subset F$ be an n-dimensional linear space, $n \geq 1$. U is called a strong (weak) oscillation space iff no nonconstant $f \in U$ has a weak (strong) oscillation of length $n + 1$.

Definition: A linear space $U \subset F$ is called 1-normed iff U contains the constant functions.

Lemma 8.2: a) Every weak oscillation space is a weak Haar space.
b) Every strong oscillation space is a Haar space and a weak oscillation space.
c) If M consists of more than n points, every strong oscillation space $U \subset F$ is 1-normed.

Proof: a) and b) are immediate from lemma 8.1 and the definitions.
c) If U does not contain the constant functions, every $f \in U$ defined by $f(t_1) = \ldots = f(t_n) = 1$ for some $(t_1,\ldots,f_n) \in \Delta_n(M)$ would have a weak oscillation of length $n + 1$.

For the following we denote by Q the set of all $f \in U$ with a strong oscillation of length n.
Analogously to theorem 7.6 we get:

Theorem 8.3: Let $U \subset F$ be an n-dimensional 1-normed weak oscillation space. Then U contains an (n-1)-dimensional 1-normed weak oscillation space.

Proof: Let $K := \{t \in M | U|_{M \cap (-\infty,t]}$ consists of the constants$\}$,

$$\tilde{M} := \begin{cases} M, & \text{if } K = \emptyset, \\ (M \setminus K) \cup \{p\} & \text{for some } p \in K, \text{ if } K \neq \emptyset, \end{cases}$$

$a := \inf \tilde{M}$ and $M_0 = \tilde{M} \setminus \{a\}$. We distinguish three cases:
1) $K \neq \emptyset$ and $t := \inf M_0 > a$:
Let $L = \{f \in U | f(a) = f(t) = 0\}$, and suppose there is an $f \in L \cap Q$. So there exist points $x_1,\ldots, x_n \in M_0$ with $x_1 < \ldots < x_n$ forming a strong oscillation of f of length n. Without loss of generality let $f(x_1) < f(x_2)$.

As $t \notin K$, there is a $g \in U$ with $g(a) \neq g(t)$, and as U contains the constants, there is an $h \in U$ with $h(a) = 1 = -h(t)$. For sufficiently small $\varepsilon > 0$, either $f + \varepsilon g \in U$ has a strong oscillation in $a, t, x_2, \ldots, x_n$, or f has a strong oscillation in $t, x_1, \ldots, x_n$. As this contradicts the hypothesis that U is a weak oscillation space, we get $L \cap Q = \emptyset$.

2) $K \neq \emptyset$ and $a = \inf M_0$:

For $x \in M_0$ let

$L_x := \{f \in U \mid f(a) = f(x) = 0\}$,

$Q_x := \{f \in U \mid f|_{M \cap [x,\infty)}$ has a strong oscillation of length n$\}$,

$B_x := U \setminus Q_x$.

By the same reasoning as was used in theorem 7.6 to show that each set A_x is open we conclude that each set Q_x is open. Let

$\tilde{Q} = \bigcup_{x \in M_0} Q_x$ and $B = U \setminus \tilde{Q} = \bigcap_{x \in M_0} B_x$.

As the restriction of U to $\{a,x\}$ has dimension 2 for every $x \in M_0$, every space L_x has dimension $n - 2$. In the same way as in case 1) the fact $L \cap Q = \emptyset$ was derived, it can be shown here that L_x and Q_x have no common element, so $L_x \subset B_x$ for $x \in M_0$. We choose a suitable sequence $\{x_\nu\}_1^\infty$ in M_0 that monotonously decreases with $\lim_{\nu\to\infty} x_\nu = a$. We apply lemma 7.5 and conclude that B contains an (n-2)-dimensional linear space L, so $L \cap \tilde{Q} = \emptyset$.

Again, we wish to show $L \cap Q = \emptyset$. Suppose the contrary, i. e., there is an $f \in L \cap (Q \setminus \tilde{Q})$. So f has a strong oscillation of length n on $\tilde{M} = \{a\} \cup M_0$, but none on M_0. This implies there is a $t \in M_0$ such that f is monotonous on $(a,t) \cap M_0$, say monotonously decreasing. From Lemma 4 follows $f(a) = 0$. Because of the oscillation property f is positive for some $u \in (a,t) \cap M_0$, and so

$$f(y) \geq f(u) > 0 \text{ for all } y \in (a,u) \cap M_0. \quad (*)$$

For $\nu = 1,2,\ldots$, let $E_\nu := \{e_\nu^1, \ldots, e_\nu^{n-2}\}$ be a basis of L_{x_ν} with $\|e_\nu^i\| = 1$ and $E_\nu \to E := \{e^1, \ldots, e^{n-2}\}$ for $\nu \to \infty$ without loss of generality, where $\| \ \|$ is a norm induced by an inner product of U (see proof of lemma 7.5) and E a basis of L.

Let $f = \sum_{i=1}^{n-2} \alpha_i e^i$ for suitable $\alpha_i \in \mathbb{R}$, and define $f_\nu \in L_{x_\nu}$ by

$f_\nu = \sum_{i=1}^{n=2} \alpha_i e_\nu^i$. We have $f_\nu \to f$ for $\nu \to \infty$, so for all $\varepsilon > 0$ there is an ν_0 such that $\|f - f_\nu\| < \varepsilon$ for all $\nu > \nu_0$.

a) If $\|f\|_{max} := \sup_{x \in M} |f(x)|$ is a norm on U, there is a constant $\gamma > 0$ with $\|h\|_{max} < \gamma \|h\|$ for all $h \in U$. But then

$$|f(x_\nu)| = |(f-f_\nu)(x_\nu)| \leq \|f-f_\nu\|_{max} < \gamma \|f-f_\nu\| < \gamma\, \varepsilon$$

for all $\nu > \nu_0$ in contradiction to (*).

b) If some element of U is unbounded on M, norm-convergence does not imply pointwise convergence. The difficulty can be circumvented, however, by some properties of the special spaces under consideration.

Definition: $x \in \bar{M}$ is called a pole of an $f \in F$ iff $f(y) \to \infty$ or $f(y) \to -\infty$ for $y \to x-$ or $y \to x+$.

Lemma 8.4: Let $U \subset F$ be an n-dimensional weak oscillation space. Then there exist at most n points in $\bar{M}$ that are poles of some $f \in U$.

Proof: Suppose there are $n + 1$ points $p_1, \ldots, p_{n+1}$ such that each p_i is a pole of some $f_i \in U$. Now assume that for a fixed $i \leq n$, the points $p_1, \ldots, p_i$ are poles of f_i. Either f_i has a pole in p_{i+1}, or $f_i + \alpha f_{i+1}$ has poles in $p_1, \ldots, p_{i+1}$ for a suitable $\alpha \in \mathbb{R}$. Replace f_{i+1} by f_i or by $f_i + \alpha f_{i+1}$. So induction over i shows that there is a $g \in U$ with poles in $p_1, \ldots, p_{n+1}$. One easily checks that g then has a strong oscillation of length $n + 1$, a contradiction.

Lemma 8.5: Let $U \subset F$ be an n-dimensional linear space, and assume that there are at most k points $p_1, \ldots, p_k \in \bar{M}$ being poles of some $f \in U$. For $\varepsilon > 0$, let

$$M_\varepsilon := M \setminus \bigcup_{i=1}^{k} (U_\varepsilon(p_i) \setminus \{p_i\}).$$

Then $\|\ \|_\varepsilon$ with $\|f\|_\varepsilon = \sup_{x \in M_\varepsilon} |f(x)|$ for $f \in U$ is a norm for all sufficiently small $\varepsilon > 0$.

Proof: Trivially $\|\ \|_\varepsilon$ is a seminorm on U. As U has dimension n, there is a set S of n points such that the restriction of U to S is an n-dimensional Haar space. For sufficiently small $\varepsilon > 0$ we have $S \subset M_\varepsilon$, and $\|f\|_\varepsilon = 0$ for some $f \in U$ yields $f|_S \equiv 0$ and so $f \equiv 0$.

We now return to the proof of theorem 8.3, case 2 b).
If a is not a pole of a function in U, we replace $\|\ \|_{max}$ by $\|\ \|_\varepsilon$ for a suitable $\varepsilon > 0$ and argue as in case 2 a).

Now assume a is a pole of a function in U. For sufficiently small $\varepsilon > 0$, f has a strong oscillation of length $n - 1$ in $M_0 \cap M_\varepsilon$, say in points $q_1 < \ldots < q_{n-1}$. The same statement then holds for all f_ν with ν sufficiently large. For such an $x_\nu < q_1$ let $y \in [a, x_\nu) \cap M_0$ be with $f_\nu(y) \neq 0$. Then either $a, y, x_\nu, q_1, \ldots, q_{n-1}$ or $a, y, q_1, \ldots, q_{n-1}$ are strong oscillations of f_ν of length greater n, a contradiction.

3) $K = \emptyset$:

We have $M_0 = \tilde{M} = M$ and $\inf M \notin M$. Let $\phi : M \to M$ be a mapping with $\phi(x) < x$ for all $x \in M$. Let $L_x = \{f \in U | f(\phi(x)) = f(x) = 0\}$, and Q_x, B_x, $\tilde{Q}$, B defined as in 2), $x \in M$. Again $\dim L_x = n - 2$ and $L_x \cap Q_x = \emptyset$ for all $x \in M$, and by lemma 7.5. B contains an (n-2)-dimensional linear space L. As $M_0 = M$, we have $Q = \tilde{Q}$, hence $L \cap Q = \emptyset$.

In all the cases 1), 2), 3) let U_{n-1} be spanned by L and the constant functions. We have $\dim U_{n-1} = n - 1$, and no $f \in U_{n-1}$ has a strong oscillation of length n.

Theorem 8.6: Let M have property (B) and $U \subset F$ an n-dimensional 1-normed Haar space which also is a weak oscillation space. Then U contains an (n-1)-dimensional subspace with the same properties.

Proof: By theorem 8.3 U contains an (n-1)-dimensional 1-normed weak oscillation space V. Suppose some $f \in V \setminus \{0\}$ had $n - 1$ zeros $x_1 < \ldots < x_{n-1}$ in M. Note that f changes sign in each zero. We distinguish four cases:

1) There are $x_0, x_n \in M$ with $x_0 < x_1$, $x_{n-1} < x_n$. From $\dim U|_{\{x_0,\ldots,x_{n-1}\}} = n$ follows $\dim V|_{\{x_0,\ldots,x_{n-1}\}} = n - 1$. Thus there is an $i \in \{1,\ldots,n-1\}$ such that $V|_{\{x_0,\ldots,x_{i-1},x_{i+1},\ldots,x_{n-1}\}}$ is an (n-1)-dimensional Haar space.

Without loss of generality, let $f(x_0) > 0$, so $(-1)^n f(x_n) < 0$.

Let $g \in V$ be defined by

$$g(x_j) = \begin{cases} (-1)^j & \text{for } j = 0, 1, \ldots, i - 1 \\ (-1)^{j+1} & \text{for } j = i + 1, \ldots, n - 1. \end{cases}$$

For sufficiently small $\varepsilon > 0$, $f + \varepsilon g \in V$ has a strong oscillation in $x_0, x_1, \ldots, x_{i-1}, \ldots, x_n$ in contradiction to the fact V is a weak oscillation space.

2) Assume $x_1 = \inf M$ and there is an $x_n \in M$ with $x_{n-1} < x_n$. Rename x_1 into x_0. Because of property (B), there is an $x_1 \in (x_0, x_2) \cap M$. Without loss of generality, let $f(x_1) < 0$, so $(-1)^n f(x_n) < 0$.

As in case 1), there is an $i \in \{0,2,\ldots,n-1\}$ such that $V|_{\{x_0,\ldots,x_{i-1},x_{i+1},\ldots,x_{n-1}\}}$ is an (n-1)-dimensional Haar space.

For $i = 0$ let $g \in V$ be defined by $g(x_j) = (-1)^j$, $j = 1,\ldots, n - 1$.

For small $\varepsilon > 0$, $f + \varepsilon g$ has a strong oscillation in $x_0, \ldots, x_{n-1}$.

For $i \in \{2,\ldots,n-1\}$, define $g \in V$ as in case 1) and argue as in case 1).

3) $\inf M < x_1$ and $x_{n-1} = \sup M$. Reverse the order of M and apply case 2).

4) $\inf M = x_1$ and $x_{n-1} = \sup M$. Rename x_1 into x_0 and x_{n-1} into x_n. Because of property (B) there exist $x_1 \in (x_0, x_2) \cap M$ and $x_{n-1} \in (x_{n-2}, x_n) \cap M$. Again let $f(x_1) < 0$, so $(-1)^{n-1} f(x_{n-1}) < 0$. As in case 1), there is an $i \in \{0,2,\dots,n\}$ such that $V|_{\{x_0,x_2,\dots,x_n\} \setminus \{x_i\}}$ is an $(n-1)$-dimensional Haar space.
For $i = 0$ let $g \in V$ be defined by

$$g(x_j) = \begin{cases} (-1)^j & \text{for } j = 2,\dots, n-1 \\ (-1)^{j+1} & \text{for } j = n. \end{cases}$$

Then $x_0, x_1, \dots, x_{n-2}, x_n$ form a strong oscillation of $f + \varepsilon g$ for sufficiently small $\varepsilon > 0$.
For $i \in \{2,\dots,n\}$ let $g \in V$ be defined by

$$g(x_j) = \begin{cases} (-1)^j & \text{for } j = 0, 2,\dots, i-1 \\ (-1)^{j+1} & \text{for } j = i+1,\dots, n. \end{cases}$$

For sufficiently small $\varepsilon > 0$, $f + \varepsilon g$ has a strong oscillation in $x_0, \dots, x_{i-1}, x_{i+1}, \dots, x_n$.

<u>Lemma 8.7:</u> Let $U_i \subset F$ be i-dimensional 1-normed spaces with $U_1 \subset \dots \subset U_n$. Let $U_1, \dots, U_{n-1}$ be Haar spaces.
a) If U_n is a Haar space, U_n is a strong oscillation space.
b) If U_n is a weak Haar space, U_n is a weak oscillation space.

<u>Proof:</u> a) <u>$n = 2$:</u> Let $f_1 \equiv 1$ be a basis of U_1, and f_1, f_2 a basis of U_2. The determinant

$$\begin{vmatrix} 1 & 1 \\ f_2(x) & f_2(y) \end{vmatrix} = f_2(y) - f_2(x)$$

has constant sign $\neq 0$ for all $(x,y) \in \Delta_2(M)$. So f_2 is strictly monotonous.
<u>$n-1 \Rightarrow n$:</u> Suppose there are $f \in U_n \setminus U_{n-1}$ and $(t_1,\dots,t_{n+1}) \in \Delta_{n+1}(M)$ with $(-1)^i [f(t_{i+1}) - f(t_i)] \geq 0$ for $i = 1,\dots, n$.
Now let $g \in U_{n-1}$ be the function that interpolates f in $t_2, \dots, t_n$. By induction hypothesis we have $(-1)[g(t_2)-g(t_1)] < 0$ and $(-1)^n[g(t_{n+1})-g(t_n)] < 0$. So $f - g$ has a weak alternation of length $n + 1$ in $t_1, \dots, t_{n+1}$ in contradiction to the fact that U_n is a Haar space.
b) By a) U_{n-1} is a strong oscillation space. Suppose there are $f \in U_n \setminus U_{n-1}$ and $(t_1,\dots,t_{n+1}) \in \Delta_{n+1}(M)$ with $(-1)^i[f(t_{i+1})-f(t_i)] > 0$ for $i = 1,\dots, n$.

For $\varepsilon > 0$ let $g_\varepsilon \in U_{n-1}$ be defined by $g_\varepsilon(t_i) = f(t_i) + \varepsilon(-1)^i$ for $i = 2,\ldots, n$. For sufficiently small $\varepsilon > 0$, $f - g_\varepsilon$ has a strong alternation of length $n + 1$ in $t_1,\ldots, t_{n+1}$ in contradiction to the hypothesis that U_n is a weak Haar space.

Lemmas 8.2 b) and 8.8 yield together with theorem 8.3:

Theorem 8.8: Let M have property (B), and $U \subset F$ be an n-dimensional 1-normed Haar space. Then the following properties are equivalent:

a) U is a strong oscillation space.

b) U is a weak oscillation space.

c) U contains i-dimensional 1-normed Haar spaces U_i, $i = 1,\ldots, n - 1$, with $U_1 \subset \ldots \subset U_{n-1} \subset U$.

d) U has a basis which is a normed Markov system (a normed "Markov basis").

Corollary 8.9: Let M have property (B), and $U \subset F$ be an n-dimensional Haar space. Then the following properties are equivalent:

a) U contains a positive function p, and $\tilde{U} := \{h \in F \mid h = f/p \text{ for an } f \in U\}$ is a strong oscillation space.

b) U contains a positive function p, and $\tilde{U} := \{h \in F \mid h = f/p \text{ for an } f \in U\}$ is a weak oscillation space.

c) U has a Markov basis.

Proof: See lemma 3.2.

Exercises: 1)* Let $U \subset F$ be an n-dimensional 1-normed Haar space that is a weak, but not a strong oscillation space. Then $n \geq 3$, and

a) if $n \geq 4$, M consists of $n + 1$ points;

b) if $n = 3$, there exist points $t_1 < t_2 < t_3 < t_4$ in M such that $M = \{t_1\} \cup (M \cap [t_2,t_3]) \cup \{t_4\}$. $t_1,\ldots, t_4$ are the only points possibly forming a weak oscillation of some $g \in U$ of length 4, and in this case $g(t_1) = g(t_2) \neq g(t_3) = g(t_4)$.

2) Let $M = \{0,1,2,3\}$ and $U = \text{span}\{f_1,f_2,f_3\} \subset F$, where $f_1 \equiv 1$, $f_2(0) = f_2(1) = f_3(1) = f_3(2) = 0$, $f_2(2) = f_2(3) = f_3(0) = f_3(3) = 1$. Show that U is a 1-normed Haar space and a weak, but not a strong oscillation space. Find i-dimensional Haar spaces U_i, $i = 1,2$, with $U_1 \subset U_2 \subset U$.

9. Periodic Haar spaces

We refer to the setting of chapter 5 and consider finite-dimensional subspaces of the linar space F_τ of τ-periodic functions defined on M. Especially, let us recall all periodic Haar or weak Haar spaces in F_τ have odd dimension.
In this situation some definitions of the previous chapter must be modified.

Definition: Let $f \in F_\tau$ and $(t_1,\ldots,t_k) \in \Delta_k(M)$. $t_1,\ldots,t_k$ are called a periodic strong (weak) oscillation of f of length k iff $t_k < t_{k+1} := t_1 + \tau$ and $(-1)^i[f(t_{i+1})-f(t_i)]$ is positive (nonnegative) for $i = 1,\ldots, k$ or negative (nonpositive) for $i = 1,\ldots, k$.

Definition: An n-dimensional linear space $U \subset F_\tau$ is called a periodic strong (weak) oscillation space iff no nonconstant $f \in U$ has a periodic weak (strong) oscillation of length n + 2, if n is even, or of length n + 1, if n is odd.

Example: Let $f_1, f_2, f_3 : \mathbb{R} \to \mathbb{R}$ be defined by $f_1(x) \equiv 1$, $f_2(x) = \sin(x)$, $f_3(x) = \cos(x)$ for all $x \in \mathbb{R}$. Then span $\{f_1,f_2\}$, span $\{f_2,f_3\}$ and span $\{f_1,f_2,f_3\}$ are periodic strong oscillation spaces in $C_{2\pi}$.

Lemma 9.1: Let $f \in F_\tau$ and $(t_1,\ldots,t_k) \in \Delta_k(M)$ with $t_k < t_1 + \tau$ be a strong (weak) oscillation of f, $k \geq 1$ odd. Then f has a periodic strong (weak) oscillation of length k-1 in $t_1,\ldots,t_k$.

Proof: Without loss of generality we assume

$$f(t_1) \underset{(\geq)}{>} f(t_2) \underset{(\leq)}{<} \cdots \underset{(\geq)}{>} f(t_{k-1}) \underset{(\leq)}{<} f(t_k).$$

In the case $f(t_{k-1}) < f(t_1)$, the points $t_{k-1} - \tau, t_1,\ldots, t_{k-2}$ form a periodic strong (weak) oscillation of f.
In the case $f(t_{k-1}) \geq f(t_1)$, we have

$$f(t_k) \underset{(\geq)}{>} f(t_{k-1}) \underset{(\geq)}{>} f(t_1) \underset{(\geq)}{>} f(t_2),$$

and $t_2,\ldots, t_k$ form a periodic strong (weak) oscillation of f.

Lemmas 8.1 and 9.1 together yield:

Lemma 9.2: Let n be odd and $U \subset F_\tau$ an n-dimensional linear space.
a) If U is a periodic weak oscillation space, U is a periodic weak Haar space.
b) If U is a periodic strong oscillation space, U is a periodic Haar space and a periodic weak oscillation space.

As in most applications periodic Haar spaces are defined on $M = \mathbb{R}$, for the rest of this chapter we shall no longer consider the most general situation but make the hypothesis that M has property (B).
Besides, we narrow our scope to periodic Haar spaces and exclude periodic weak Haar spaces and their subspaces from our considerations. Also excluded are investigations of the structure of periodic weak and strong oscillation spaces under additional hypotheses. We are sure, though, that new results may be expected in this direction.

Lemma 9.3: Let M have property (B) and $U \subset F_\tau$ be an n-dimensional 1-normed periodic Haar space and a periodic weak oscillation space. Then U is a periodic strong oscillation space.

Proof: Suppose the contrary, i. e., a nonconstant $f \in U$ exists with a periodic weak oscillation of length $n + 1$ in points $t_1 < \ldots < t_{n+1}$. As U is a Haar space and M has property (B) we can without loss of generality assume there is exactly one i with $f(t_i) = \max\{f(t_j) \mid j=1,\ldots,n+1\}$.
Let $g \in U$ be defined by

$$g(t_j) = (-1)^{j-i} \text{ for } j = 1,\ldots, i-1, i+1,\ldots, n+1.$$

For sufficiently small $\varepsilon > 0$, $f + \varepsilon g \in U$ has a periodic strong oscillation of length $n + 1$ in $t_1,\ldots, t_{n+1}$ in contradiction to the hypothesis.

Lemma 9.4: Let M have property (B) and $U \subset F_\tau$ be an n-dimensional periodic Haar space. Then every $f \in U \setminus \{0\}$ with $n - 2$ or $n - 1$ zeros has a periodic strong alternation of length $n - 1$.

Proof: If an $f \in U \setminus \{0\}$ has zeros $t_1 < \ldots < t_{n-1}$ with $t_{n-1} < t_n := t_1 + \tau$, all points $u_1,\ldots, u_{n-1}$ with $u_i \in M \cap (t_{i-1},t_i)$ for $i = 2,\ldots, n$ form periodic strong oscillations of f because of exercise 1 in chapter 3.
Now let $f \in U \setminus \{0\}$ have $n - 2$ zeros $t_1 < \ldots < t_{n-2}$ with $t_{n-2} < t_{n-1} := t_1 + \tau$. Let $p_i \in (t_i,t_{i+1})$, $i = 1,\ldots, n - 2$, be arbitrarily fixed.
For $n = 3$ the points t_1, p_1 form a periodic strong oscillation of f of length $n - 1$.
Now assume $n \geq 5$. One easily checks that for $k = 2,\ldots, n - 2$ the set $S := \{t_1,p_1,\ldots,t_k,p_k\}$ contains $k + 1$ points forming an (ordinary) strong oscillation of f if we have $f(p_i) \cdot f(p_{i+1}) < 0$ for $i = 1,\ldots, k - 1$. Otherwise, S contains at least $k + 2$ points forming a

strong oscillation of f. So the set $\{t_1,p_1,\ldots,t_{n-2},p_{n-2}\}$ contains n points forming a strong oscillation of f of length n, or we have $f(p_i)\cdot f(p_{i+1}) < 0$ for $i = 1,\ldots, n-3$. In this case, however, we have $f(p_{n-2})\cdot f(p_1) > 0$, and the set $\{t_2,p_2,\ldots,t_{n-2},p_{n-2},t_1+\tau,p_1+\tau\}$ contains n points forming a strong oscillation of f of length n. In either case the statement now follows from lemma 9.1.

<u>Theorem 9.5:</u> Let M have property (B) and $U \subset F_\tau$ be an n-dimensional periodic (1-normed) strong oscillation space, $n \geq 3$ odd. Then U contains an (n-2)-dimensional periodic strong oscillation space.

<u>Proof:</u> Let $Q := \{f\in U \mid f$ has a periodic strong oscillation of length $n-1\}$ throughout this proof. Let $\{x_k\}_{k=1}^{\infty}$ be a strictly decreasing, convergent sequence in M, say $x_k \to 0$ for $k \to \infty$. For $k = 1,2,\ldots$ let

$L_k := \{f\in U \mid f(x_k)=f(x_{k+1})=f(x_{k+2})=0\}$,

$Q_k := \{f\in U \mid f|_{M\cap[x_k,\tau)}$ has a periodic strong oscillation of length $n-1\}$,

$B_k := U \setminus Q_k$.

Further, let $\tilde{Q} := \bigcup_{k=1}^{\infty} Q_k$ and $B := U \setminus \tilde{Q} = \bigcap_{k=1}^{\infty} B_k$.

Q_k is open (see proof of theorem 7.6), B_k is closed and L_k is an (n-3)-dimensional linear space for every k.

Suppose for some k the sets L_k and Q_k have a common element f. If $t_1,\ldots, t_{n-1} \in M \cap [x_k,\tau)$ is a periodic strong oscillation of f of length $n-1$, it is not difficult to see that the set $\{x_{k+2},x_{k+1},x_k,t_1,\ldots,t_{n-1}\}$ contains a periodic weak oscillation of f of length $n+1$ in contradiction to the hypothesis. So we get $L_k \subset B_k$ for all k.

By lemma 7.5 B contains an (n-3)-dimensional linear space L, so $L \cap \tilde{Q} = \emptyset$.

a) If $\lim_{k\to\infty} x_k = 0$ is no point of M, we have $\tilde{Q} = Q$, so $L \cap Q = \emptyset$.

b) If $0 \in M$ we suppose there is an $f \in L \cap (Q\setminus\tilde{Q})$. So let $0, t_2,\ldots, t_{n-1}$ be a periodic strong oscillation of f of length n. Lemma 7.5 implies $f(0) = 0$ since we have $f \in L$. Without loss of generality assume $f(t_2) > 0$. As in part 2) of the proof of theorem 8.3, a contradiction can be derived. So we get $L \cap Q = \emptyset$ in this case, too.

If we define $U_{n-2} := \operatorname{span}\{L,1\}$, where 1 is the constant function ($1\in U$ because of lemma 8.2 c), we get $U_{n-2} \cap Q = \emptyset$. So U_{n-2} is an (n-2)-dimensional normed periodic weak oscillation space. By lemma 9.2 b) U is a periodic Haar space, and lemma 9.4 yields that U_{n-2} is a periodic Haar space. By lemma 9.3 U_{n-2} is a periodic strong oscillation space.

It seems unlikely that every normed periodic Haar space should be a weak oscillation space, even in C_T with $M = \mathbb{R}$. So theorem 9.5 probably does not comprise all normed periodic Haar spaces. The statement "Every n-dimensional periodic Haar space contains an (n-2)-dimensional periodic Haar space" is until now neither proved nor disproved. It is also unknown which cases may be reduced to theorem 9.5 by division by a suitable positive function.

10. Examples

In this chapter we give various examples of Haar or weak Haar spaces to show that some of the central results of the preceding chapters, namely theorems 6.5, 7.7, are "sharp" insofar as they do not hold in general if one of the hypotheses is dropped.

Example 10.1: Let M be the half-open interval $[-1,1)$, n even and $U := \text{span}\{f_1,\dots,f_n\}$, where

$$f_1(t) = t,$$
$$f_i(t) = t^{i-2}(t^2-1) \text{ for } i = 2,\dots,n, \quad t \in M.$$

Every $f \in U \setminus \{0\}$ may be written

$$f = \alpha f_1 - h,$$

with $\alpha \in \mathbb{R}$, $h(t) = (t^2-1)\,P(t)$, P a real polynomial of degree $\le n-2$.
1) For $\deg(P) \le n-3$, f has at most n-1 zeros on $\mathbb{R}$. For $\deg(P) = n-2$ we have

$$\lim_{t\to\infty} h(t) = \lim_{t\to-\infty} h(t) = \pm\,\infty.$$

From $h(1) = h(-1) = 0$ follows that for every $\alpha \in \mathbb{R}$ there is a point $u \in \mathbb{R} \setminus M$ with $\alpha f_1(u) = h(u)$. So f has at most n-1 zeros on M. We conclude - see theorem 3.5 - that U is a Haar space.
2) For $\alpha = 0$ we have $f(-1) = 0$.
For $\alpha \neq 0$ we have $f(1) = -f(-1) = \alpha$, and f has a zero in $(-1,1)$. So U contains no one-dimensional Haar subspace and consequently has no Markov basis. We see that in theorem 7.7, the condition that M contains neither its infimum nor its supremum cannot be dropped in general.
3) In the terminology of theorem 6.5, $\alpha \neq 0$ is equivalent to $a < s_1$. One easily checks that for $\alpha \neq 0$, the set SZ(f) consists of oddly many points. Thus the statement of theorem 6.5 is never true for the space U under consideration if $a < s_1$ and $n - (k + 2l)$ is even.

Example 10.2: Let $M: = [0,2\pi)$, $n \geq 3$ odd and U the space of trigonometric polynomials of degree $\leq \frac{n-1}{2}$. By example 5.6, U is a Haar space.

1) U contains no Haar subspaces of even dimension (see corollary 5.2) and so has no Markov basis.

2) The periodicity of the functions in U implies that for any $f \in U$ with $f(0) \neq 0$ - i.e., $a < s_1$ - SZ(f) consists of an even number of points. Again, the statement of theorem 6.5 never holds if $n - (k+2l)$ is even.

We now turn to closed domains of definition.

Example 10.3: Let $M = [-1,1]$, n odd and $V: = \text{span}\,\{p_1,\dots,p_n\}$, where

$$p_i(t) = (1-t)\, f_i(t), \quad i = 1,\dots,n-1,$$

$$p_n(t) = 1,$$

for $t \in M$, where $f_1,\dots,f_{n-1}$ are the functions from example 10.1. Every $p \in V \setminus \{0\}$ may be written $p = g - h$, where

$$g(t) = \gamma(t^2-t) + \delta,$$
$$h(t) = (t-1)^2(t+1)\;P(t)$$

with $\gamma, \delta \in \mathbb{R}$ and P a real polynomial of degree $\leq n-3$.

1) For $\deg(P) \leq n-4$, p has at most n-1 zeros on $\mathbb{R}$. Now assume $\deg(P) = n-3$ and without loss of generality let

$$\lim_{t\to\infty} h(t) = \infty, \text{ so } \lim_{t\to-\infty} h(t) = -\infty.$$

As h is of higher degree than g, we have $h(t) > g(t)$ for all t sufficiently large and $h(t) < g(t)$ for all t sufficiently small. For $g(-1) < 0 = h(-1)$ or $g(1) > 0 = h(1)$, g and h must have at least one common point in $\mathbb{R} \setminus M$, and p has at most n-1 zeros on M.

Now let $g(-1) \geq 0 \geq g(1)$, so $\gamma \geq 0 \geq \delta$. We have

$$g'(t) = \gamma(2t - 1),$$
$$h'(t) = (t-1)\;Q(t),$$

where Q is a real polynomial of degree n-2. From $h'(1) = 0 \leq g'(1)$ and $h'(t) > g'(t)$ for all t sufficiently large follows that p' has a zero in $[1,\infty)$ and so at most n-2 zeros in $[-1,1)$. Then p has at most n-1 zeros on M. By theorem 3.5 V is a Haar space.

2) We suppose V contains an (n-1)-dimensional Haar space X on M.

Then the restriction of $Y: = \{f \in X | f(1) = 0\}$ to $[-1,1)$ is an $(n-2)$-dimensional Haar space. By theorem 6.5 Y contains a function r that is positive on $[-1,1)$. On the other hand we have

$$r(t) = (1-t)\ s(t)$$

for a suitable $s \in U$, where U is the space from example 10.1. But then s has a zero in $[-1,1)$, and we arrive at a contradiction.

3) If, using the terminology of theorem 6.5, we wish to construct an $f \in V$ with $a = -1 < s_1$ and $s_k = b = 1$, we are led to example 10.1, part 3). So again in this case the statement of theorem 6.5 is never true if $n - (k+2l)$ is even.

Example 10.4: Let $M = [0,\pi]$, $n = 2k$ and $U: = \text{span}\{f_1,g_1,\dots,f_k,g_k\}$, where

$$f_\nu(t) = \cos(\nu t)$$
$$g_\nu(t) = \sin(\nu t), \nu = 1,\dots,k,\ t \in M.$$

As in example 5.6, every $f \in U$ may be written in the form

$$f(x) = z^{-k}\ P(z),\quad z = e^{ix},$$

where P is a complex polynomial of degree $\leq 2k$.

1) As U does not contain the constants, we get

$$P(z) = Q(z) + z^{k+1}\ R(z),$$

where Q and R are complex polynomials of degree $\leq k-1$. Hence 0 is a zero of the k-th derivative of P. Repeated application of LUCAS' theorem (see, e. g., OBRESCHKOFF 1966, p. 8/9) shows that 0 must be contained in the convex hull $H(P^{(\nu)})$ of the zeros of $P^{(\nu)}$ for every $\nu \in \{0,1,\dots,k\}$.

If an $f \in U$ had exactly 2k zeros $t_1,\dots,t_{2k}$ in $[0,\pi]$, $t_1<\dots<t_{2k}$, P would have exactly 2k zeros $z_1,\dots,z_{2k} \in \mathbb{C}$ with $\arg(z_i) = t_i$ for all i. So 0 would lie on the border of H(P). Since $z_1,\dots,z_{2k}$ are simple zeros, again by LUCAS'theorem the zeros of P' lie in the interior of H(P), so $0 \in H(P') \subset \text{int}(H(P))$. So no $f \in U \setminus \{0\}$ has more than 2k-1 zeros, and U is a Haar space.

2) We suppose there are i-dimensional Haar spaces U_i, $i = 1,\dots,2k$, with $U_1 \subset \dots \subset U_{2k} = U$. Let $g \in U$ be a positive function with $U_1 = \text{span}\{g\}$, and

$$\tilde{U}_i: = \{f | f = \frac{h}{g} \text{ for an } h \in U_i\}$$

for all i. By theorem 8.8 $\tilde{U}_{2k}$ is a strong oscillation space.
Now let S be a complex polynomial of degree 2k with simple, pairwise distinct zeros $y_1,\ldots,y_{2k}$ on the unit circle and $y_1 = 1, y_2 = -1$, $y_{2\nu-1} = \overline{y_{2\nu}}$ for $\nu = 2,\ldots,k$. A simple computation shows that S is of the form $S(z) = Q_o(z) + z^{k+1}R_o(z)$ with suitable polynomials Q_o and R_o of degree $\leq k-1$, and so S corresponds to a function $h \in U$ with 2k distinct zeros in $t_i = \arg(z_i)$, $i = 1,\ldots,2k$ (exercise 1.).
The preceding argument shows that especially for every $\varepsilon \in (0,\pi)$ there is an $h_\varepsilon \in U \setminus \{0\}$ with zeros $t_1,\ldots,t_{2k}$ with

$$0 = t_1 < \ldots < t_{2k-1} < \varepsilon < \pi < t_{2k} < \pi + \varepsilon.$$

Furthermore, let $\max_{x \in M} |h_\varepsilon(x)| = 1$. If x_ε denotes the extremum of h_ε in (t_{2k-1}, t_{2k}), by continuity reasons we have $|h_\varepsilon(x_\varepsilon)| = 1$ and $x_\varepsilon < \pi$ for all sufficiently small $\varepsilon > 0$, and $\lim_{\varepsilon \to 0} h_\varepsilon(\pi) = 0$. This means that h_ε has a strong oscillation of length n+1 on M for all sufficiently small $\varepsilon > 0$.
Being a continuous function on a compact set, g attains its minimum $\gamma > 0$. Because of $\lim_{\varepsilon \to 0} h_\varepsilon(\pi) = 0$, for sufficiently small $\varepsilon > 0$ the function $\frac{h_\varepsilon}{g} \in \tilde{U}$ has a strong oscillation of length n+1, a contradiction. So we have shown U has no Markov basis, but it remains an open question if U contains an (n-1)-dimensional Haar space.

Considering the previous examples, it seems reasonable to conjecture that the zeros of a function f in a Haar space can arbitrarily be prescribed in the sense of theorem 6.5 if and only U has a Markov basis, a statement which, if true, would make much of chapter 6 superfluous. However, while the "if"-part is immediate from theorem 6.5 a), the "only if"-part does not hold in general. As an example, take example 10.4 with k = 2. As we have seen, U has no Markov basis. On the other hand, for every pair $x_1, x_2 \in [0,\pi]$ with $x_1 \neq x_2$ there is an $f \in U$ with $Z(f) = \{x_1,x_2\}$ and $DZ(f) = \emptyset$, and a $g \in U$ with $Z(f) = DZ(f) = \{x_1\}$. The possibility of prescribing 0,1 or 3 zeros plus double zeros for a function in U follows from theorem 6.5 a) and c) (see also exercise 2.).
The following example demonstrates that parts d) and e) of theorem 6.5 do not hold in general if M fails to have property (B).

<u>Example 10.5:</u> Let $a,b \in \mathbb{R}$ with $a < -1$, $1 \leq b$ be arbitraily fixed, and $M: = \{a\} \cup [-1,1) \cup \{b\}$. Let $f_1,\ldots,f_4 \in F$ be defined by the following table giving the values of $f_i(x)$:

i	x = a	x ∈ [-1,1)	x = b
1	1	1	0
2	0	x	0
3	1	x^2	0
4	-4	x^3	1

We define linear spaces by

$W: = \text{span}\{f_1,\dots,f_4\}$, $\hat{W}: = \{f \in W \mid f(b) = 0\} = \text{span}\{f_1,f_2,f_3\}$,
$V: = \hat{W}|_{M\setminus\{b\}}$, $\hat{V}: = \{f \in \hat{W} \mid f(a) = 0\} = \text{span}\{f_2,f_1-f_3\}$,
$U: = \hat{V}|_{[-1,1)}$.

1) U is the space from example 10.1 for n=2.
2) V is a 3-dimensional Haar space.

<u>Proof:</u> Suppose an $f \in V\setminus\{0\}$ has a weak alternation of length 4 in $\{a\} \cup [-1,1)$. So f has two distinct zeros in [-1,1), say p and q. Without loss of generality we have

$$f(x) = (x-p)(x-q) \text{ for } x \in [-1,1)$$

and so

$$f = f_3 - (p+q)\, f_2 + pq\, f_1.$$

But then $f(a) = 1 + pq > 0$, and f has no alternation of length 4, a contradiction.
3) W is a 4-dimensional Haar space.
<u>Proof:</u> Suppose an $f \in W \setminus \{0\}$ has a weak alternation of length 5 in M. $f(b) = 0$ is impossible, because then $f \in V$, and f cannot have a weak alternation of length 4 in $M \setminus \{b\}$.
Without loss of generality assume $f(b) > 0$. We first look at the case that f has exactly two distinct zeros p and q in [-1,1). As $f|_{[-1,1)}$ is a polynomial of degree 3, this polynomial has another zero $r \in \mathbb{R} \setminus [-1,1)$. As we have $f(b) > 0$ and f has a weak alternation of length 5 on M, we get $r \geq 1$.
Without loss of generality we have

$$f(x) = (x-p)(x-q)(x-r) \text{ for } x \in [-1,1),$$

and so

$$f = f_4 - (p+q+r)f_3 + (pq+pr+qr)f_2 - pqr\, f_1.$$

But then

$$f(a) = -4 - (p+q+r) - pqr \leq 4 - q - pq \leq -2,$$

and from $f(-1) \leq 0$ follows that f cannot have a weak alternation of length 5.

If f has three distinct zeros p,q,r in $[-1,1)$, we get

$$f(a) = -4 - (p+q+r) - pqr < 0,$$

and again f has no weak alternation of length 5 because of $f(-1) \leq 0$.

4) As U contains no function without a zero, there is no $f \in V$ with $Z(f) = \{a\}$ and we have a counterexample to theorem 6.5 e), and there is no $g \in W$ with $Z(g) = \{a,b\}$, a counterexample to theorem 6.5 d).

We close this section with an open problem:
Under what conditions does an n-dimensional Haar space contain an (n-2)-dimensional Haar subspace?

Exercises: 1) Give the details of example 10.4, part 2).
2) Consider example 10.4 with k=2. Find functions $f \in U$ with

a) $Z(f) = SZ(f) = \{0,\pi\}$,

b) $Z(f) = DZ(f) = \{\frac{\pi}{2}\}$,

c) $Z(f) = \emptyset$.

3) *Consider lemma 7.3, assume $f_n \equiv 1$ and put $\tilde{\phi} := (f_1,\ldots,f_{n-1})$.
a) Show that lemma 7.3 a),b),c) are then equivalent to the following statement:
For every pair $a,b \in \mathbb{R}^{n-1}$ the (n-2)-dimensional affine space

$$A(a,b) := \{z \in \mathbb{R}^{n-1} \mid \langle z,a\rangle = b\}$$

intersects $\tilde{\phi}(M)$ in at most n-1 points.
b) Show that U is a generalized Haar space that contains no generalized Haar space of dimension n-1 if and only if a) and the following statement hold:
For every $b \in \mathbb{R}^{n-1}$ there is an $a \in \mathbb{R}^{n-1}$ and for every $a \in \mathbb{R}^{n-1}$ there is a $b \in \mathbb{R}^{n-1}$ such that A(a,b) intersects $\tilde{\phi}(M)$ in n-1 points.

(Hint: Look at n=3 first).

4) *Let U: = span$\{f_1,f_2,f_3\}$ with $f_3 \equiv 1$. Using exercise 3 b), show that in the following examples U is a 3-dimensional Haar space that contains no 2-dimensional Haar space.

a) $M = [-\pi,\pi]$, $f_1(t) = \cos t$, $f_2(t) = \begin{cases} \sin t & \text{for } t \in [-\pi,\frac{\pi}{2}] \\ (1+\sin t)/2 & \text{for } t \in (\frac{\pi}{2},\pi] \end{cases}$

(VOLKOV 1958, KIEFER and WOLFOWITZ 1965).

b) $M = [0,\pi]$, $f_1(t) = h(t) \cdot \sin t$, $f_2(t) = h(t) \cdot \cos t$,

$h(t) = \begin{cases} \sin t & \text{for } t \in [0,\frac{\pi}{2}] \\ 1 & \text{for } t \in (\frac{\pi}{2},\pi] \end{cases}$ (NEMETH 1966).

c) Same as b), but $h(t) = t$ (ZIELKE 1971, 1973).

d) Same as b), but $M = [0,\frac{3}{2}\pi]$ and $h(t) = \sqrt{2}-\cos t$ (HADELER 1973).

5) Find more examples analogous to those of exercise 4).

6)* Consider example 10.3 with n=3. Prove the following statements:

a) W is a 2-dimensional Haar subspace of $V_{|[-1,1)}$ if and only if $W = \text{span}\{p_1+\alpha p_3, p_2\}$ for some $\alpha \in [0,2)$.

b) $U_{|[-1,1)}$ has no Markov basis.

(This example shows that extendability of the domain M by one point does not guarantee the existence of a Markov basis; compare with corollary 7.2 c).)

7) Let $M = [-1,1]$, $n \geq 3$ odd and U: = span$\{f_1,\ldots,f_n\} \subset C(M)$ be defined by $f_1(x) \equiv 1$,

$$f_2(x) = \frac{1}{2}x^3 - x,$$

$$f_j(x) = x^{j-3}(x^2-1)^2 \text{ for } j = 3,\ldots,n, \quad x \in M.$$

Show that U is a Haar space containing no Haar space of even dimension and there is no $f \in U\setminus\{0\}$ with $Z(f) = \{1\}$ (HAVERKAMP 1978).

IV. Analytical properties

While in the previous chapters we mainly considered linear spaces without regard of a special basis the emphasis will now be shifted a bit towards single functions. We shall usually be dealing with normed Markov systems or normed weak Markov systems.

11. Differentiability

For a normed (weak) Markov system $f_1,\dots, f_n \in F$ (for the definition, see chapter 8) we shall in the following denote by U_i the span of $\{f_1,\dots,f_i\}$, $i = 1,\dots, n$.

Lemma 11.1: If $f_1,\dots, f_n \in F$ is a normed Markov system every $f \in U_n$ is bounded on $M \cap [a,b]$ where $a, b \in M$ are arbitrary.

Lemma 11.1 says in other words that every $f \in U_n$ has poles at most in $a = \inf M$ or $b = \sup M$, and this only if $a \notin M$ or $b \notin M$, respectively.

Proof of Lemma 11.1: For $n = 1$ the statement is trivial.
$n - 1 \Rightarrow n$: Suppose there is an $f \in U_n \setminus U_{n-1}$ and points $a, b \in M$ such that f is unbounded on $M \cap [a,b]$. By lemma 8.7 a) U_n is a strong oscillation space. Thus f must possess a pole $d \in \bar{M} \cap [a,b]$ and without loss of generality f be strictly increasing and unbounded on some set $M \cap [c,d)$ for a suitably chosen $c \in M \cap (-\infty,d)$.
Let $(t_1,\dots,t_{n-1}) \in \Delta_{n-1}(M)$ with $c \le t_1$, $t_{n-1} < d$ be fixed, and let $g \in U_{n-1}$ be defined by $g(t_i) = (-1)^{n-i}$, $i = 1,\dots, n-1$. By lemma 8.7 a) we have $g(x) < -1$ for all $x \in M \cap (t_{n-1},\infty)$, and by the induction hypothesis g is bounded on $M \cap [c,d]$. If we let $t_{n+1} \in M \cap [d,b]$ be fixed, for sufficiently small $\varepsilon > 0$ we have
$\operatorname{sign}(g + \varepsilon f)(t_i) = (-1)^{n-i}$ for $i = 1,\dots, n-1, n+1$.
On the other hand, for every such ε the function $g + \varepsilon f \in U_n$ is unbounded from above on $M \cap (t_{n-1},d)$. So there is a $t_n \in M \cap (t_{n-1},t_{n+1})$ with $(g + \varepsilon f)(t_n) > 0$. So $g + \varepsilon f$ has a strong alternation of length $n + 1$ in $t_1,\dots, t_{n+1}$ in contradiction to the fact that U_n is a Haar space.

As another application of lemma 8.7 a) we get:

Theorem 11.2: Let $f_1,\dots, f_n \in F$ be a normed Markov system, $a := \inf M \notin M$ and $b := \sup M \notin M$. Assume that for all $i = 1,\dots, n$ there exists $\alpha_i := \lim_{\substack{x\to a+\\ x\in M}} f_i(x)$ as well as $\beta_i := \lim_{\substack{x\to b-\\ x\in M}} f_i(x)$. Let

$\overline{f}_i : \{a,b\} \cup M \to \mathbb{R}$ be defined by $\overline{f}_i(a) = \alpha_i$, $\overline{f}_i(b) = \beta_i$ and $\overline{f}_i|_M = f_i$. Then $\overline{f}_1,\ldots, \overline{f}_n$ form a normed Markov system.

Proof: Let $\bar{U}_i := \text{span}\ \{\overline{f}_1,\ldots,\overline{f}_i\}$, $i = 1,\ldots, n$. We only consider the case $i = n$. Suppose an $\bar{f} \in \bar{U}_n \setminus \{0\}$ has n zeros $x_1 <\ldots< x_n$ in $M \cup \{a,b\}$. As U_n is a Haar space, either $x_1 = a$ or $x_n = b$ holds, say $x_1 = a$. Let $p \in M \cap (a,x_2)$ be a point with $f(p) \neq 0$, and $q \in M \cap (a,p)$ be so near to a that $|f(q)| < |f(p)|$.
If $x_n < b$, the points $q, p, x_2,\ldots, x_n$ form a weak oscillation of f of length $n + 1$ in contradiction to lemma 8.7 a).
If $x_n = b$, we choose points $r, s \in M \cap (x_{n-1},b)$ with $r < s$ and $0 \neq |f(r)| > |f(s)|$. One checks that the set $\{q,p,x_2,\ldots,x_{n-1},r,s\}$ contains at least $n + 1$ consecutive points forming a weak oscillation of f of length $n + 1$, a contradiction.

We shall subsequently carry over the term "derivative" to normed Markov systems. This will turn out to be a proper generalization of the classical Markov system formed by the power functions $f_i(x) = x^{i-1}$. In this context it will become apparent that the concept of non-continuous Čebyšev systems is more than an artificial extension of the usual continuous cases, because here discontinuous Haar spaces arise quite naturally even if one deals with continuous Haar spaces in the beginning.

Definition: Let $a \in M$ and $f \in F$. Then f is said to have the right-hand M-limit α in a (written $\alpha = M - \lim\limits_{x\to a+} f(x)$) iff for every $\varepsilon > 0$ there exists an $y \in M \cap (a,\infty)$ with $|f(x)-\alpha| < \varepsilon$ for all $x \in M \cap (a,y]$. The limits $M - \lim\limits_{x\to a+} \inf f(x)$ and $M - \lim\limits_{x\to a+} \sup f(x)$ are defined analogously, as well as the corresponding left-hand limits.

Obviously if for some $a, b \in M$ with $a < b$ the set $M \cap (a,b)$ is empty, we have $f(b) = M - \lim\limits_{x\to a+} f(x)$ and $f(a) = M - \lim\limits_{x\to b-} f(x)$.

Theorem 11.3: Let M be a set containing neither its infimum nor its supremum, and $f_1,\ldots, f_n \in F$ a weak normed Markov system, $n \geq 3$, and $f_1,\ldots, f_{n-1}$ a Markov system. Then the following statements hold:
a) A linear operator $D_+ : U_n \to F$ is defined by

$$(D_+f)(a) := M - \lim_{x\to a+} \frac{f(x)-f(a)}{f_2(x)-f_2(a)}$$

for $f \in U_n$ and $a \in M$.

b) $D_+ f_2, \ldots, D_+ f_n$ form a normed weak Markov system.

c) If M has property (B) and $f_1, \ldots, f_n$ is a Markov system, $D_+ f_2, \ldots, D_+ f_n$ form a normed Markov system.

Proof: We have $D_+ f_1 \equiv 0$ and $D_+ f_2 \doteq 1$. So all the statements a), b), c) are fulfilled for $n - 1 = 2$.

$n - 1 \Rightarrow n \geq 3$: a) Let $a \in M$ and $f \in U_n \setminus U_{n-1}$ be fixed. If there is a $y \in M \cap (a,\infty)$ with $M \cap (a,y) = \emptyset$, we have

$$(D_+f)(a) = \frac{f(y)-f(a)}{f_2(y)-f_2(a)} .$$

Now assume $a = \inf\{M \cap (a,\infty)\}$. Without loss of generality we may set $f(a) = f_2(a) = 0$ and assume f_2 is strictly increasing and $f \geq 0$ on $M \cap (a,c)$ for a suitably chosen $c \in M \cap (a,\infty)$. We first suppose

$$M - \lim_{x \to a+} \sup \frac{f(x)}{f_2(x)} = \infty.$$

By induction hypothesis for every $g \in U_{n-1} \setminus \{0\}$ there exists $M - \lim\limits_{x \to a+} \frac{g(x)}{f_2(x)}$. For all $g \in U_{n-1} \setminus \{0\}$ with $g(a) = 0$ and $x \in M \cap (a,\infty)$ with $g(x) \neq 0$ we have $\frac{f(x)}{f_2(x)} = \frac{f(x)}{g(x)} \cdot \frac{g(x)}{f_2(x)}$. Hence follows

$$M - \lim_{x \to a+} \sup \frac{f(x)}{g(x)} = \infty$$

for all $g \in U_{n-1} \setminus \{0\}$ with $g(a) = 0$ that are strictly increasing on $M \cap [a,\infty)$.

Now let $(t_1, \ldots, t_{n-1}, t_{n+1}) \in \Delta_n(M)$ with $t_{n-1} = a$ and $g \in U_{n-1}$ be defined by

$$g(t_i) = \begin{cases} (-1)^{n-1-i} & \text{for } i = 1, \ldots, n-2 \\ 0 & \text{for } i = n-1. \end{cases}$$

By lemma 8.7 a), g is strictly increasing on $M \cap [a,\infty)$, so $g(t_{n+1}) > 0$. For a sufficiently small $\varepsilon > 0$, we have sign $(g - \varepsilon f)(t_i) =$ sign $g(t_i)$ for $i = 1, \ldots, n-1, n+1$. Besides, we see that

$$M - \lim_{x \to a+} \inf \frac{(g-\varepsilon f)(x)}{g(x)} = -\infty,$$

so $(g - \varepsilon f)(t_n) < 0$ for some $t_n \in M \cap (a,t_{n+1})$. Thus $g - \varepsilon f \in U_n$ has a strong oscillation of length $n + 1$ in $t_1, \ldots, t_{n+1}$ in contradiction to lemma 8.7 b) by which U_n is a weak oscillation space.

Now suppose we have

$$\beta := M - \lim_{x \to a+} \inf \frac{f(x)}{f_2(x)} < \gamma := M - \lim_{x \to a+} \sup \frac{f(x)}{f_2(x)} .$$

Set $\delta := (\beta+\gamma)/2$. So there exist strictly decreasing sequences $t_1, t_2, \ldots \in M$ and $u_1, u_2, \ldots \in M$ with the same limit and

$$\lim_{i\to\infty} \frac{f(t_i)}{f_2(t_i)} = \beta < \delta < \gamma = \lim_{i\to\infty} \frac{f(u_i)}{f_2(u_i)} .$$

For almost all i we then have $(f - \delta f_2)(t_i) < 0 < (f - \delta f_2)(u_i)$. By considering suitable subsequences we may without loss of generality assume $t_1 > u_1 > t_2 > u_2 > \ldots$. So $f - \delta f_2 \in U$ has a strong alternation of arbitrary length, a contradiction.

$D_+ : U_n \to F$ obviously is a linear operator.

b) From $\text{kern}(D_+) = U_1 = \text{span}\{f_1\}$ follows $\dim D_+ U_n = \dim U_n - \dim \text{kern}\,(D_+) = n - 1$, and $D_+ f_2, \ldots, D_+ f_n$ are linearly independent.

Suppose there are $f \in U_n$ and $(t_1,\ldots,t_n) \in \Delta_n(M)$ with $(-1)^i (D_+f)(t_i) > 0$ for $i = 1,\ldots, n$. Then there exist $u_1,\ldots, u_n \in M$ with $t_1 < u_1 \le t_2 < u_2 \le \ldots \le t_n < u_n$ and

$$(-1)^i \frac{f(u_i)-f(t_i)}{f_2(u_i)-f_2(t_i)} > 0 \text{ for } i = 1,\ldots, n,$$

so $(-1)^i[f(u_i)-f(t_i)] > 0$ for $i = 1,\ldots, n$. Then the set $\{t_1,u_1,\ldots,t_n,u_n\}$ contains at least $n + 1$ points forming a strong oscillation of f in contradiction to lemma 8.7 b). So $D_+ U_n$ is a weak Haar space.

c) By theorem 4.5 now $D_+ U_n$ is a Haar space, or else there is a derivative $D_+ f \in D_+ U_n \setminus \{0\}$ with infinitely many zeros. As f is not constant, we may choose $n - 1$ zeros $t_1,\ldots, t_{n-1}$ of $D_+ f$ with $t_1 < \ldots < t_{n-1}$ and such that $f|_{\{t_1,\ldots,t_{n-1}\}}$ is not constant.

Let $g \in U_{n-1}$ be the function interpolating f in $t_1,\ldots, t_{n-1}$. We have $(D_+(g-f))(t_i) = (D_+g)(t_i)$ for $i = 1,\ldots, n - 1$ and $D_+ g \in U_{n-1}$. As $g - f$ changes sign strongly in each of the t_i's, $D_+(g-f)$ has a weak alternation of length $n - 1$ in $t_1,\ldots, t_{n-1}$, so the same holds for $D_+ g$. Since we have $D_+ g \neq 0$ the induction hypothesis that $D_+ U_{n-1}$ is an $(n-2)$-dimensional Haar space leads to a contradiction.

<u>Corollary 11.4:</u> Under the hypotheses of theorem 11.3 the following statements hold:

a) A linear operator $D_- : U_n \to F$ is defined by

$$(D_-f)(a) := M - \lim_{x\to a-} \frac{f(x)-f(a)}{f_2(x)-f_2(a)}$$

for $f \in U_n$ and $a \in M$.

b) $D_- f_2, \ldots, D_- f_n$ form a normed weak Markov system.
c) If M has property (B) and $f_1, \ldots, f_n$ is a Markov system, $D_+ f_2, \ldots, D_+ f_n$ form a normed Markov system.

Statement c) of theorem 11.3 does not hold in general when the hypothesis that M has property (B) is dropped (this fact was overlooked in ZIELKE (1974)). We demonstrate this by an example:

Example 11.5: Let $M := (-\infty,-1] \cup [1,\infty)$ and define $f_1, f_2, f_3 \in F$ by

$$f_1(x) \equiv 1,$$
$$f_2(x) = x \text{ for } x \in M,$$
$$f_3(x) = \begin{cases} (x+1)^2 & \text{for } x \le -1 \\ (x-1)^2 & \text{for } x \ge 1 \end{cases}$$

f_1, f_2, f_3 form a normed Markov system, but $D_+ f_3$ has 1 and -1 as zeros.

If M contains its infimum or supremum, the differentiability statement of theorem 11.3 does not necessarily hold in these points. Let, for example, $M := [-1,1]$, f_1 and f_2 defined as in example 11.5 and $f_3(x) = \sqrt{1-x^2}$ for $x \in M$. f_1, f_2, f_3 form a normed Markov system but $D_+ f_3$ is not defined in -1.

Exercise: Give a detailed proof of corollary 11.4 without using theorem 11.3.

12. Properties of the derivatives

In this chapter we make the general assumption that $f_1, \ldots, f_n \in M$ is a normed Markov system. To make sure that $D_+ f_2, \ldots, D_+ f_n$ and $D_- f_2, \ldots, D_- f_n$ are normed Markov systems we would need the hypotheses: M has property (B), $a := \inf M \notin M$, $b := \sup M \notin M$ (theorem 11.3). Instead, we make the more restrictive assumption that M is an open interval (a,b). This is done to avoid rather lengthy notations which would not offer much, if any additional understanding. The generalizations to sets M with property (B) are left to the reader.
The first lemma is trivial:

Lemma 12.1: $U \subset C_{\underset{(-)}{+}}(M)$ if and only if $f_2 \in C_{\underset{(-)}{+}}(M)$.

<u>Lemma 12.2:</u> If we have $U \subset C_{\overset{+}{(-)}}(M)$, then also $D_{\overset{+}{(-)}} U \subset C_{\pm}(M)$.

<u>Proof:</u> By lemma 12.1, it is sufficient to show $D_+ f_3 \in C_+(M)$ if $U \subset C_+(M)$. Let $t \in M$ be fixed, and, without loss of generality, assume $f_2(t) = f_3(t) = (D_+f_3)(t) = 0$, f_2 and D_+f_3 strictly increasing, and suppose $\lim_{x \to t+} (D_+f_3)(x) = 1$.

As $(D_+f_3)(x)$ is positive for $x > t$, f_3 is increasing for $x > t$, and clearly strictly increasing, too. Thus $(f_3(x)/f_2(x))$ is positive for all $x > t$. As we have

$$0 = (D_+f_3)(t) = \lim_{x \to t+} (f_3(x)/f_2(x)),$$

for every $\varepsilon' > 0$ there is an $\varepsilon \in (0,\varepsilon')$ and a $z \in (t,b)$ with $(f_3(z) / f_2(z)) = \varepsilon$. As $f_3(x) \geq \varepsilon f_2(x)$ for all $x \in (t,z)$ is ruled out, we have $f_3(x) < \varepsilon f_2(x)$ for all $x \in (t,z)$, because otherwise $f_3 - \varepsilon f_2$ would change sign weakly in more than three points in $[t,z]$. Now let $y \in (t,z)$ with $f_2(y) < \varepsilon f_2(z)$, and define $r := \inf \{s \in (y,b) \mid ((f_3(s) - f_3(y))/(f_2(s) - f_2(y))) < \frac{1}{2}\}$.
If we had $r \geq z$, we would get
$\varepsilon f_2(z) = f_3(z) > f_3(z) - f_3(y) \geq \frac{1}{2}(f_2(z) - f_2(y)) \geq \frac{1}{2}(1-\varepsilon) f_2(z)$,
which is impossible for $\varepsilon < \varepsilon' < \frac{1}{3}$. Thus we have $r < z$. From $(D_+f_3)(y) > 1$ follows $y < r$. Then there exist points $l \in (y,r)$ and $s \in (r,z)$ with

$$\frac{f_3(s) - f_3(y)}{f_2(s) - f_2(y)} < \frac{1}{2} < \frac{f_3(l) - f_3(y)}{f_2(l) - f_2(y)}, \qquad (*)$$

and so

$$\begin{aligned} f_3(s) - f_3(l) &= (f_3(s) - f_3(y)) - (f_3(l) - f_3(y)) \\ &< \tfrac{1}{2}(f_2(s) - f_2(y)) - \tfrac{1}{2}(f_2(l) - f_2(y)) = \tfrac{1}{2}(f_2(s) - f_2(l)) \end{aligned} \qquad (**)$$

Defining $g \in U$ by $g(x) = f_3(l) - \frac{1}{2}(f_2(l) - f_2(x))$ for $x \in M$, we get

$f_3(l) = g(l)$,
$f_3(y) < g(y)$ from $(*)$,
$f_3(s) < g(s)$ from $(**)$.

Thus $f_3 - g$ changes sign weakly in y, l and s, which implies $f_3(x) < g(x)$ for all $x \in M \cap (-\infty,l)$. But then

$$0 = f_3(t) < g(t) = f_3(1) - \frac{1}{2} f_2(1) < (\varepsilon - \frac{1}{2}) f_2(1),$$

which is impossible for $\varepsilon < \varepsilon' < \frac{1}{2}$.

As the functions in U, D_+ U or D_- U are piecewise monotone and bounded (see lemmas 8.7 and 11.1), they are discontinuous in at most countably many points.
In the following three results, connections between D_+ U and D_- U are established.

Lemma 12.3: For every $f \in U$ and $x \in M$ we have

$$\lim_{y\to x-} (D_-f)(y) = \lim_{y\to x-} (D_+f)(y), \quad \lim_{y\to x+} (D_-f)(y) = \lim_{y\to x+} (D_+f)(y).$$

Proof: Suppose, for example, that for some $x \in M$ and $f \in U$ we had

$$\beta := \lim_{y\to x-} (D_-f)(y) = \lim_{y\to x-} \lim_{\varepsilon\to 0+} \frac{f(y) - f(y-\varepsilon)}{f_2(y) - f_2(y-\varepsilon)}$$

and

$$\gamma := \lim_{y\to x-} (D_+f)(y) = \lim_{y\to x-} \lim_{\varepsilon\to 0+} \frac{f(y+\varepsilon) - f(y)}{f_2(y+\varepsilon) - f_2(y)} \text{ with } \beta < \gamma.$$

Then there is an $y_0 \in M$, $y_0 < x$, such that

$$\lim_{\varepsilon\to 0+} \frac{f(z) - f(z-\varepsilon)}{f_2(z) - f_2(z-\varepsilon)} < \frac{\beta+\gamma}{2} < \lim_{\varepsilon\to 0+} \frac{f(z+\varepsilon) - f(z)}{f_2(z+\varepsilon) - f_2(z)} \text{ for all } z \in (y_0,x).$$

Hence for each $z \in (y_0,x)$ there is an $\varepsilon(z) > 0$ such that

$$\frac{f(z) - f(l)}{f_2(z) - f_2(l)} < \frac{\beta+\gamma}{2} < \frac{f(r) - f(z)}{f_2(r) - f_2(z)}$$

for all $l \in (z-\varepsilon(z),z)$ and $r \in (z,z+\varepsilon(z))$.
For $g := f - ((\beta+\gamma)/2) f_2$ we get $g(l) > g(z) < g(r)$; so g has a local minimum in z. As this holds for every $z \in (y_0,x)$, g is constant, and we get a contradiction.
For $\beta > \gamma$, consider $-f$ instead of f. For right-hand limits the proof is analogous.

It might be suspected that, beyond lemma 12,3, one has $(D_-f)(x)$, $(D_+f)(x) \in [\liminf_{y\to x} (D_\pm f)(y), \limsup_{y\to x} (D_\pm f)(y)]$. This is not true, as the following example shows:

<u>Example 12.4:</u> Let a normed Markov system $f_0, f_1, f_2, \ldots \in \mathbb{R}^{\mathbb{R}}$ be defined by

$$f_i(x) = \begin{cases} (x-1)^i & \text{for } x \le 0 \\ (x+1)^i & \text{for } x > 0, \ i = 0,1,2,\ldots \end{cases} .$$

One has $\lim_{x\to 0} (D_\pm f_3)(x) = 3$, but $(D_+f_3)(0) = 1$.

<u>Lemma 12.5:</u> Let $p \in M$ and $\alpha_i \in [\liminf_{y\to p} f_i(y), \limsup_{y\to p} f_i(y)]$ be arbitrarily fixed, and $\tilde{f}_1, \ldots, \tilde{f}_n \in \mathbb{R}^M$ be defined by

$$\tilde{f}_i(x) = \begin{cases} f_i(x) & \text{for } x \in M\setminus\{p\} \\ \alpha_i & \text{for } x = p. \end{cases}$$

Then $\tilde{f}_1, \ldots, \tilde{f}_n$ form a normed Markov system.

<u>Proof:</u> Let $\tilde{f} = \sum_{i=1}^{n} \gamma_i \tilde{f}_i \neq 0$ and $f = \sum_{i=1}^{n} \gamma_i f_i$. We have $\tilde{f}(p) \in [\liminf_{y\to p} f(y), \limsup_{y\to p} f(y)]$. If $\tilde{f}$ had a strong alternation of length n+1, one of the alternation points would be p, and thus f would have a strong alternation of length n+1, a contradiction. If $\tilde{f}$ had n zeros, one of these would be p, and f would change sign strongly in the remaining n-1 zeros. So f would have constant non-zero sign near p, say f positive near p. From $\tilde{f}(p) = 0$ follows that for sufficiently small $\varepsilon > 0$, $f - \varepsilon f_1$ has a strong alternation of length n+2, a contradiction.

Lemma 12.5 immediately yields:

<u>Theorem 12.6:</u> For i = 1,...,n, let $\tilde{f}_i \in \mathbb{R}^M$ be defined such that $\tilde{f}_i(x)$ is arbitrarily fixed in $[\liminf_{y\to x} f_i(y), \limsup_{y\to x} f_i(x)]$ for $x \in M$. Then $\tilde{f}_1, \ldots, \tilde{f}_n$ form a normed Markov system.

13. Two applications

1. Let $M \subset \mathbb{R}$ be an open interval, $n \geq 1$ and $r \in C^n(M)$ with $0 \not\equiv r^{(n)} \geq 0$. Let $\mathbb{P}_k$ be the Haar space of real polynomials of degree k, $k = 0,1,\ldots$ (see exercise 1.1 and theorem 3.5).

Theorem 13.1: Let $V := \operatorname{span}\{r, \mathbb{P}_{n-1}\} \subset F$. Then

a) V is an $(n+1)$-dimensional weak Haar space.

b) No $h \in V$ has more than n separated zeros.

c) If an $h \in V$ has $n+1$ zeros $x_1,\ldots,x_{n+1} \in M$ with $x_1<\ldots<x_{n+1}$, h vanishes on $[x_1,x_{n+1}]$.

Proof: b): The assertion follows from theorem 4.3 because M consists entirely of essential points.

a) We begin with an auxilliary result of some independent interest:

Lemma 13.2: Let $U \subset C_+(M)$ be an n-dimensional (weak) Haar space and $g \in C(M)$ strictly monotone. Let $c \in M$ be arbitrarily fixed. Then

$$V := \{h \in C(M) \mid h(x) = \int_c^x f(t)\, dg(t) + \gamma \text{ for an } f \in U, \gamma \in \mathbb{R} \text{ and all } x \in M\}$$

is an $(n+1)$-dimensional (weak) Haar space.

Proof: Let $h \in V$ with $h(x) = \int_c^x f(t)\, dg(t) + \gamma$ for an $f \in U$ and $\gamma \in \mathbb{R}$.

If there were $x_1,\ldots,x_{n+2} \in M$ with $x_1<\ldots<x_{n+2}$ and $\operatorname{sign} h(x_i) = (-1)^i$ for all i, for $i = 1,\ldots,n+1$ we would get

$$0 < (-1)^i[h(x_{i+1}) - h(x_i)] = (-1)^i \int_{x_i}^{x_{i+1}} f(t)\, dg(t).$$

But then each interval $[x_i,x_{i+1}]$, $i = 1,\ldots,n+1$, would contain a point y_i with $\operatorname{sign} f(y_i) = (-1)^i$ in contradiction to the hypothesis. Now let U be a Haar space. If h had $n+1$ zeros $z_1,\ldots,z_{n+1} \in M$ with $z_1<\ldots<z_{n+1}$, we would get

$$0 = \int_{z_i}^{z_{i+1}} f(t)\, dg(t) \quad \text{for } i = 1,\ldots,n.$$

So f would have a zero in each interval (z_i,z_{i+1}), $i = 1,\ldots,n$, and thus vanish identically, implying $h \equiv 0$.

We now return to the proof of theorem 13.1 a). Let f_i, $i = 1,\ldots,n$, be defined by $f_i(x)=x^{i-1}$ for $x \in M$ and all i. The synopsis of theorem 11.3 b) and Lemma 13.2 shows that all of the following statements (A_j), $j = 0,1,\ldots,n-1$, are equivalent:

(A_j): $f_1,\ldots,f_{n-j}$, $r^{(j)}$ form a weak Markov system.

(A_{n-1}) is in turn equivalent to the fact that $r^{(n-1)}$ is nonconstant and monotone , i.e., $r^{(n)}$ has weakly constant sign and $r^{(n)} \not\equiv 0$,which is guaranteed by the hypotheses. So (A_o) holds and implies a).

c) For <u>n = 1</u> the assertion is trivial.

<u>n-1 => n</u>: If we had $h_{|[x_1,x_{n+1}]} \not\equiv 0$ there would exist $t_i \in (x_i,x_{i+1})$ with $h'(t_i) = 0$ for $i = 1,\ldots,n$ and $h'_{|[t_1,t_n]} \not\equiv 0$ in contradiction to the induction hypothesis.

2. In this section we shall characterize the Haar subspaces of the space $\mathbb{P}$ of real polynomials on $M = \mathbb{R}$.

<u>Theorem 13.3:</u> Let $U \subset \mathbb{P}$ be an n-dimensional linear space containing the constants. Then the following statements are equivalent:

a) U is a Haar space.

b) U has a Markov basis, every Markov basis $r_1,\ldots,r_n$ of U is normed, and $\deg r_{i+1} - \deg r_i$ is positive and odd for $i = 1,\ldots,n-1$.

<u>Proof:</u> We shall prove the assertion for rational functions rather than for polynomials. In this context, a rational function $r = \frac{p}{q}$, where p,q are polynomials, $p \not\equiv 0$, has degree $\deg r = \deg p - \deg q$.

We make use of the following auxiliary results on rational functions the proofs of which are left to the reader:

<u>Proposition A:</u> If r and s are rational functions, we have $\deg (r/s) = \deg r - \deg s$.

<u>Proposition B:</u> If r is a rational function with $\deg r \neq 0$, for the derivative r' we have $\deg r' = \deg r - 1$.

<u>Proposition C:</u> If r is a nonconstant rational function with $\deg r = 0$, we have $\deg r' \leq -2$.

The implication b) => a) is trivial.

<u>a) => b):</u> By theorem 7.7 U has a Markov basis $r_1,\ldots,r_n$.

The case <u>n = 1</u> is trivial.

<u>n-1 => n:</u> As $r_1,\ldots,r_{n-1}$ is a normed Markov system by the induction

hypothesis, r_2 is strictly monotone on $\mathbb{R}$. So deg r_2 is odd, and moreover, deg $r_2 > 0$, for otherwise we would have $\lim_{x\to-\infty} r_2(x) = \lim_{x\to\infty} r_2(x)$.

By theorem 11.3 c) the rational functions $D_+r_2,\ldots,D_+r_n$ form a normed Markov system. By induction hypothesis the differences deg D_+r_{i+1} - deg D_+r_i are positive and odd for $i = 2,\ldots,n-1$. As $D_+r_i = r_i'/r_2'$ for all i, we have that $\deg(r_{i+1}'/r_2') - \deg(r_i'/r_2')$ is positive and odd for $i = 2,\ldots,n-1$, and so deg r_{i+1}' - deg r_i' is positive and odd for $i = 2,\ldots,n-1$ by proposition A.
With proposition B follows

$$0 \le \deg r_2 - 1 = \deg r_2' < \deg r_3' < \ldots < \deg r_n'.$$

So each $r_i, i \ge 2$, has positive degree by propositions C and B, and

$$\deg r_{i+1} - \deg r_i = \deg r_{i+1}' - \deg r_i' \text{ for } i = 2,\ldots,n-1.$$

Corollary 13.4: Let $f_1,\ldots,f_n \in \mathbb{P}$ be defined by $f_i(x) = x^{\alpha_i}$, $i = 1,\ldots,n$, where $0 = \alpha_1 < \alpha_2 < \ldots < \alpha_n$. Then $U := \text{span}\{f_1,\ldots,f_n\}$ is a Haar space if and only if $\alpha_{i+1} - \alpha_i$ is odd for $i = 1,2,\ldots,n-1$.

Proof: While the necessity is immediate from theorem 13.3, the sufficiency may be derived from Descartes' rule (see, e.g., SCHMEISSER and SCHIRMEIER 1976, p. 58/59). Alternatively, and more in line with the ideas of this chapter, one may use induction via Lemma 13.2, i.e., define $g = f_2$ and let U be spanned by functions u_i with $u_i(x) = x^{\alpha_i-\alpha_2}$, $i = 2,\ldots,n$.

Theorem 13.5: Let $U \subset \mathbb{P}$ be an n-dimensional linear space. Then the following statements are equivalent:

a) U is a Haar space.

b) U has a Markov basis, and for every Markov basis $r_1,\ldots,r_n$ of U deg r_1 is even and deg r_{i+1} - deg r_i is positive and odd for $i = 1,\ldots,n-1$.

Proof: The part b) => a) again is trivial.
a) => b): By theorem 7.7 U has Markov basis $r_1,\ldots,r_n$ and deg r_1 obviously is even. Division by r_1 leads to a situation where theorem 13.3 is applicable. So $\deg \frac{r_{i+1}}{r_1} - \deg \frac{r_i}{r_1} = \deg r_{i+1} - \deg r_i$ is positive and odd for $i = 1,\ldots,n-1$.

V. Adjoined functions

In this section we shall deal with the existence of functions adjoined to a weak Haar or Haar space which are defined as follows:

Definition: Let $U \subset F$ be an n-dimensional (weak) Haar space. $f \in F$ is called (weakly) adjoined to U if and only if span$\{U,f\}$ is an (n+1)-dimensional (weak) Haar space.

We shall not deal with generalized Haar spaces because no nontrivial results seem to be known in this context.
In chapter 14 we shall derive existence theorems using a fairly direct approach. A more involved approach is prepared in chapters 15 and 16 and carried out in chapter 17. Nevertheless, the material of chapters 15 and 16 is of considerable independent interest.

14. Adjoined functions in the discontinuous case

Throughout this chapter we shall assume that M has property (B) and contains neither its infimum nor supremum. Our main result is:

Theorem 14.1: Let $f_1,\ldots,f_n \in F$ be a normed Markov system. Then there exists a function adjoined to span$\{f_1,\ldots,f_n\}$.

It is useful to break the proof into several parts.

Lemma 14.2: Let $g \subset F$ be strictly monotone, $U \in F$ an n-dimensional linear space, and $V := \{h : g(M) \to \mathbb{R} \mid h = f \circ g^{-1}$ for an $f \in U\}$.
Then the following statements hold:

a) M has property (B) if and only if g(M) has property (B).
b) U is a Haar space if and only if V is a Haar space.
c) A function $p: g(M) \to \mathbb{R}$ is adjoined to V if and only if $p \circ g \in F$ is adjoined to U.
d) If M is an interval and $g \in C(M)$, g(M) is an interval and g is a homeomorphism.

Lemma 14.2 is a minor extension of exercise 3 in chapter 3.
If we apply lemma 14.2 to the situation of theorem 14.1 with $g = f_2$ we see that we only need to consider the case: $f_2(t) = t$ for $t \in M$.

Lemma 14.3: Let the hypotheses of theorem 14.1 obtain, assume $f_2(t) = t$ for all $t \in M$, and let $a := \inf M$, $b := \sup M$, $I := (a,b)$.

Then there exist weak Haar spaces $\overline{U}_i \subset C(I)$, $i = 1,\dots,n$, with the following properties:

a) $\overline{U}_{i|_M} = U_i$ for $i = 1,\dots,n$.

b) For every $\overline{f} \in \overline{U}_n$ and $x \in I$ there exists $(D_+\overline{f})(x) := \lim\limits_{y\to x+} \frac{\overline{f}(y)-\overline{f}(x)}{y-x}$.

d) $D_+\overline{U}_i$ is an $(i-1)$-dimensional weak Haar space with

$D_+\overline{U}_{i|_M} = D_+U_i$, $i = 1,\dots,n$.

d) For every $\overline{f} \in \overline{U}_n$, $D_+\overline{f}$ is continuous from the right.

Proof: a) We shall construct $\overline{f}_1,\dots,\overline{f}_i \in C(I)$ forming a normed weak Markov basis of $\overline{U}_i$, $i = 1,\dots,n$.

For $n = 2$ we define $\overline{f}_2 \in C(I)$ by $\overline{f}_2(t) = t$ for all $t \in I$.

$n-1 \Rightarrow n$: We set $f := f_n$. First we show that f may be continuously extended to $I \cap \overline{M}$. Let $x \in I \cap (\overline{M}\setminus M)$ be fixed. By lemma 8.7 a), I can be split into $k \le n-1$ subintervals $A_1,\dots,A_k$ such that f is strictly monotone on each of the sets $M \cap A_i$, $i = 1,\dots,k$. As we have $a,b \notin M$, there exist $\ell := \sup\{y \in M | y < x\}$ and $r := \inf\{y \in M | x < y\}$.

f is bounded near ℓ and r because of lemma 11.1, and so there exist the limits

$$f_-(\ell) := M\text{-}\lim_{y\to\ell-} f(y) \text{ and } f_+(r) := M\text{-}\lim_{y\to r+} f(y).$$

For $\ell = x < r$ we define $\overline{f}(x) := f_-(\ell)$, for $\ell < x = r$ we define $\overline{f}(x) = f_+(r)$. For $\ell = x = r$ it remains to show $f_-(x) = f_+(x)$.

Suppose the contrary, say $0 = f_-(x) < f_+(x)$ without loss of generality. Let $t_1,\dots,t_{n-2},t_{n+1} \in M$ be points with $t_1<\dots<t_{n-2}<x<t_{n+1}$. For every $\overline{h}_o \in \overline{U}_{n-1}\setminus\{0\}$ with zeros in $t_1,\dots,t_{n-2}$ we have $\overline{h}_o(x) \neq 0$ because of lemma 8.7 a) and $x \in \overline{M}$. Now let $\overline{h} \in \overline{U}_{n-1}$ be a function with $h(t_i) = (-1)^{n-1-i}$ for $i = 1,\dots,n-2$, and

$\overline{g} := \overline{h}_o(x) \cdot \overline{h} - \overline{h}(x) \cdot \overline{h}_o \in U_{n-1}$.

So we have

$$g(t_i) = (-1)^{n-1-i} \text{ for } i = 1,\dots,n-2,$$
$$\overline{g}(x) = 0,$$
$$\overline{g}(y) > 0 \text{ for } y > x.$$

Then there is an $\alpha > 0$ such that

$$(-1)^{n-1-i}(g-\alpha f)(t_i) > 0 \text{ for } i = 1,\dots,n-2,n+1,$$
$$(\overline{g}-\alpha f_-)(x) = 0$$

As we have $0 = \overline{g}(x) < f_+(x)$, there is a $t_n \in M \cap (x,t_{n+1})$ with

$(g-\alpha f)(t_n) < 0$. Besides, there is a $t_{n-1} \in M \cap (t_{n-2},x)$ with $(g-\alpha f)(t_{n-2}) < (g-\alpha f)(t_{n-1}) > (g-\alpha f)(t_n)$. But then $g - \alpha f$ has a strong oscillation of length n+1 in $t_1,\ldots,t_{n+1}$ in contradiction to lemma 8.7 a).

We now fill the "gaps" of $\overline{M}$ by linear functions, i.e., for $t \in I\setminus\overline{M}$ we put

$$\overline{f}(t) = \overline{f}(t_-) + \frac{t-t_-}{t_+-t_-}(\overline{f}(t_+) - \overline{f}(t)),$$

where $t_-: = \max\{u \in \overline{M} | u < t\}$, $t_+: = \min\{u \in \overline{M} | t < u\}$.

It is easy to see that if some $\overline{q} \in \overline{U}_n$ had a strong alternation of length n+1, the same would hold for $q \in U_n$.

b) and c): The properties stated are direct consequences of the construction and theorem 11.3.

d) Let $x \in I$ be fixed. If x lies in an interval of $I\setminus\overline{M}$ or is the left endpoint of such an interval, the statement follows from the construction of $\overline{f}$. If x is the limit of a decreasing sequence of points in M, a slight modification of lemma 12.2 gives the statement.

<u>Lemma 14.4:</u> Let the situation of lemma 14.3 obtain, $A \subset I$ be a compact subinterval and $\overline{f} \in \overline{U}_n$. Then the difference quotient $q(x,y): = \dfrac{\overline{f}(y)-\overline{f}(x)}{y-x}$ is bounded for $(x,y) \in A$, $x \neq y$.

<u>Proof:</u> <u>n = 2:</u> The difference quotient is a constant.

<u>n-1 => n:</u> Suppose there are sequences $x_1,x_2,\ldots$ and $y_1,y_2,\ldots$ in A and an $\overline{f} \in \overline{U}_n$ with $x_k < y_k$ for all k and $|q(x_k,y_k)| \to \infty$ for $k \to \infty$. It is sufficient to consider pairs (x_k,y_k) with $x_k,y_k \in \overline{M}$, because if $\ell,r,\tilde{\ell},\tilde{r} \in \overline{M}$ are points with $\ell \leq x_k \leq r$ and $\tilde{\ell} \leq y_k \leq \tilde{r}$ and such that $\overline{f}$ is linear on $[\ell,r]$ and on $[\tilde{\ell},\tilde{r}]$, one easily checks

$$|q(x_k,y_k)| \leq \max\{|q(\ell,r)|,|q(\tilde{\ell},\tilde{r})|,|q(\ell,\tilde{\ell})|,|q(\ell,\tilde{r})|,|q(r,\tilde{\ell})|,|q(r,\tilde{r})|\}.$$

As for every k there exist $\tilde{x}_k,\tilde{y}_k \in M$ with $|q(\tilde{x}_k,\tilde{y}_k)| > |q(x_k,y_k)|-1$ we may without loss of generality assume $x_k,y_k \in A \cap M$ for all k. As A is compact we may assume $x_k \to x \in A \cap \overline{M}$ and $y_k \to y \in A \cap \overline{M}$ for $k \to \infty$. As $|\overline{f}(y_k) - \overline{f}(x_k)|$ is bounded because of the construction of $\overline{f}$ and lemma 11.1, $y_k - x_k$ goes to 0 for $k \to \infty$, and we have $x = y$. The reasoning is now similar to parts of the proof of lemma 14.3: Without loss of generality let $\overline{f}(x) = 0$ and $q(x_k,y_k) \to \infty$ for $k \to \infty$. As in the previous proof construct a function $\overline{g} \in \overline{U}_{n-1}$ with $g(t_i) = (-1)^{n-1-i}$ for $i = 1,\ldots,n-2$, $\overline{g}(x) = 0$, $\overline{g}(y) > 0$ for $y > x$

where $t_1,\dots,t_{n-2} \in M$ with $t_1<\dots<t_{n-2} < x$ are arbitrarily fixed. Choose $t_{n+1} \in M$ with $x < t_{n+1}$. Again there is an $\alpha > 0$ such that

$$(-1)^{n-1-i}(g-\alpha f)(t_i) > 0 \text{ for } i = 1,\dots,n-2,n+1 \text{ and } (\bar{g}-\alpha f)(x) = 0.$$

From lemma 8.7 we infer that $\bar{g} - \alpha\bar{f}$ is increasing on $[x-\varepsilon,x+\varepsilon]$ for a suitable $\varepsilon > 0$, and so

$$(g-\alpha f)(t_{n-2}) < (g-\alpha f)(x_k) < (g-\alpha f)(y_k) < (g-\alpha f)(t_{n+1})$$

for all k sufficiently large. On the other hand, the induction hypothesis implies

$$\frac{(g-\alpha f)(y_k)-(g-\alpha f)(x_k)}{y_k - x_k} = \frac{g(y_k)-g(x_k)}{y_k - x_k} - \alpha q(x_k,y_k) < 0$$

for all k sufficiently large, and $y_k - x_k > 0$ gives a contradiction.

As every $\bar{f} \in \bar{U}_n$ is Lipschitz-bounded on every compact subinterval of I by the preceding lemma, we get:

<u>Lemma 14.5:</u> Under the hypotheses of lemma 14.3, every $\bar{f} \in \bar{U}_n$ is absolutely continuous on every compact subset of I.

<u>Proof of theorem 14.1:</u> As we may put $f_2(t) = t$ for $t \in M$, the hypotheses of lemma 14.3 are fulfilled.
For <u>n = 1</u> take $g \in F$ with $g(t) = t$ as the adjoined function, for <u>n = 2</u> take $g \in F$ with $g(t) = t^2$.
<u>n-1 => n:</u> By induction hypothesis there is a function $r \in F$ adjoined to $D_+ U_n$. As in the proof of lemma 14.3, r can be extended continuously to $\bar{r}$ defined on $I \cap \bar{M}$. The "gaps", however, are filled in a different way: For $f \in I\setminus\bar{M}$, define

$$\bar{r}(t) = \bar{r}(t_-), \text{ where } t_-: = \max\{u \in \bar{M} \quad u < t\}.$$

Clearly, $W: = \operatorname{span}\{D_+\bar{U}_n,r\}$ is an n-dimensional weak Haar space, and $W_{|M} = \operatorname{span}\{D_+U,r\}$ an n-dimensional Haar space.
For a linear space $L \subset F$ of integrable functions, we define

$$I(L): = \Big\{h \in C(I)\,|\,h(x) = \int_c^x w(t)\,dt + \alpha \text{ for some } w \in L,$$
$$c \in I,\ \alpha \in \mathbb{R} \text{ and all } x \in I\Big\}.$$

From lemma 13.2 we know $I(W)$ is an $(n+1)$-dimensional weak Haar space, and lemma 14.5 yields $\bar{U}_n = I(D_+\bar{U}_n)$. In order to show that for $g \in I(W)$ with

$$g(x) = \int_c^x \bar{r}(t)\,dt \quad \text{for some fixed } c \in I,$$

the restriction $g_{|M}$ is adjoined to U_n, we must prove $U_{n+1} = I(W)|_M$ is an (n+1)-dimensional Haar space. Suppose the contrary, i.e., there is an $h \in I(W)\setminus\{0\}$ with zeros $t_1,\dots,t_{n+1} \in M$, $t_1<\dots<t_{n+1}$. Then there exist a nonconstant $w \in W$, $c \in I$ and $\alpha \in \mathbb{R}$ with

$$\int_c^{t_i} w(t)\,dt = -\alpha \quad \text{for } i = 1,\dots,n+1,$$

and, consequently,

$$\int_{t_i}^{t_{i+1}} w(t)\,dt = 0 \quad \text{for } i = 1,\dots,n.$$

As M has property (B), every interval (t_i,t_{i+1}) contains infinitely many points of M, and w cannot vanish on $M \cap (t_i,t_{i+1})$, $i = 1,\dots,n$. Since w is continuous from the right, w must assume positive and negative values on each (t_i,t_{i+1}), and so has a strong alternation of length n+1 in contradiction to the fact that W is an n-dimensional weak Haar space.

Theorem 14.6: Let M be an open interval and $f_1,\dots,f_n \in C(M)$ a normed Markov system. Then there is an $f \in C(M)$ adjoined to $U_n := \text{span}\{f_1,\dots,f_n\}$.

Proof: In the proof of theorem 14.1 the adjoined function g was continuous in case $f_2(t) = t$ for all $t \in M$. Otherwise, a transformation $f_2^{-1}: f_2(M) \to M$ was necessary, i.e., the final adjoined function was $g \circ f_2$ (see lemma 14.2 c)). Under our hypotheses $g \circ f_2$ is continuous.

Theorem 7.7 and lemma 3.2 imply:

Corollary 14.7: Let $U \subset F$ be an n-dimensional Haar space. Then there is an $f \in F$ adjoined to U. If, in addition, M is an (open) interval and $U \subset C(M)$, there is an $f \in C(M)$ adjoined to U.

Exercises: 1) Prove lemma 14.2. 2) Why is it sufficient for the proof of lemma 14.4 to consider points in M?

15. Haar spaces of differentiable functions

In this chapter we shall make the general hypothesis that M is an interval. For a Markov system $f_0,\dots,f_n$ we shall, differently from chapter 11, define $V_i = \operatorname{span}\{f_0,\dots,f_i\}$, $i = 0,1,\dots,n$. From theorem 11.3 we immediately infer:

Theorem 15.1: Let M be open and $f_1,\dots,f_n \in C^1(M)$ a normed Markov system such that f_2' has constant sign on M. Then

$$(D_+f)(x) = (D_-f)(x) = \frac{f'(x)}{f_2'(x)}$$

holds for all $f \in \operatorname{span}\{f_1,\dots,f_n\}$ and $x \in M$, and $1, \frac{f_3'}{f_2'},\dots,\frac{f_n'}{f_2'}$ form a normed Markov system.

If $f_1,\dots,f_n \in C^1(M)$ is a normed Markov system, f_2' has weakly constant sign in any case. Without the additional hypothesis on f_2 in theorem 15.1, however, even the statement $D_+f = D_-f$ may be violated, as the following example shows.

Example: Let a normed Markov system $f_1,f_2,f_3 \in C^1(\mathbb{R})$ be defined by

$$f_2(x) = \begin{cases} -x^2 & \text{for } x < 0 \\ x^2 & \text{for } x \geq 0 \end{cases}, \quad f_3(x) = \begin{cases} -x^3 & \text{for } x < 0 \\ x^3+x^2 & \text{for } x \geq 0. \end{cases}$$

Considering theorem 11.3 and lemma 13.2, one easily checks that f_1,f_2,f_3 in fact span a Haar space. One has $(D_-f_3)(0) = 0$, $(D_+f_3)(0) = 1$.

Repeated application of theorem 15.1 yields:

Theorem 15.2: Let M be open and $f_0,\dots,f_n \in C^k(M)$ a Markov system with $k \leq n$. Then there exist functions $f_i^{[\nu]} \in C^{k-\nu}(M)$ for $i = 1,\dots,n$ and $\nu = 0,1,\dots,k$ that are defined recursively by

$$\left.\begin{aligned} f_i^{[0]}(x) &= \frac{f_i(x)}{f_0(x)} \quad \text{for } i = 0,1,\dots,n, \\ f_i^{[\nu+1]}(x) &= \frac{\left(f_i^{[\nu]}\right)'(x)}{\left(f_{\nu+1}^{[\nu]}\right)'} \quad \text{for } \nu = 1,\dots,k \text{ and } i = 0,1,\dots,n \end{aligned}\right\} \quad (*)$$

for $x \in M$ if each $f_{\nu+1}^{[\nu]}$, $\nu = 0,1,\dots,n-1$, has constant sign on M.

Furthermore, $f_\nu^{[\nu]},\ldots,f_n^{[\nu]}$ form normed Markov systems for $\nu = 0,1,\ldots k$.

Definition: A Markov system $f_o^{[o]},\ldots,f_n^{[o]} \in C^k(M)$ is called an extended Markov system of order k if and only if $\left(f_{\nu+1}^{[\nu]}\right)'$, as defined by (*), has constant sign on M for $\nu = 0,1,\ldots,k-1$. An extended Markov system of order n is simply called extended Markov system.

In theorem 15.2 we have $f_i^{[\nu]} \equiv 0$ for $i < \nu$. So for $k = n$ and an extended Markov system $f_o,\ldots,f_n$ we get a triangular scheme

$$\begin{array}{ccccc} f_o^{[o]} & f_1^{[o]} & f_2^{[o]} & \ldots & f_n^{[o]} \\ & f_1^{[1]} & f_2^{[1]} & \ldots & f_n^{[1]} \\ & & f_2^{[2]} & \ldots & f_n^{[2]} \\ & & & \ddots & \vdots \\ & & & & f_n^{[n]} \end{array}$$

where each row is an extended normed Markov system. Setting $w_o = f_o$ and $w_{\nu+1} = f_{\nu+1}^{[\nu]}$ for $\nu = 0,1,\ldots,n-1$, for arbitrary $c \in M$ we get a basis $h_o,h_1,\ldots,h_n$ of V_n that has the following integral representation:

$$h_o(x) = w_o(x)$$

$$h_1(x) = w_o(x) \int_c^x w_1(t_1)\, dt_1,$$

$$h_2(x) = w_o(x) \int_c^x w_1(t_1) \int_c^{t_1} w_2(t_2)\, dt_2\, dt_1,$$

$$\vdots$$

$$h_n(x) = w_o(x) \int_c^x w_1(t_1) \int_c^{t_1} w_2(t_2) \ldots \int_c^{t_{n-1}} w_n(t_n) dt_n \ldots dt_2 dt_1.$$

Moreover, one easily verifies (exercise 3):

Lemma 15.3: Let M be open and $f_o,\ldots,f_n \in C^k(M)$ an extended Markov system of order $k \le n$, $p \in M$, $f \in V_n$ and $m \in \{1,\ldots,k\}$. Then one has

$$f^{[o]}(p) = f^{[1]}(p) = \ldots = f^{[m-1]}(p) = 0 \neq f^{[m]}(p)$$

if and only if $f(p) = f'(p) = \ldots = f^{(m-1)}(p) = 0 \neq f^{(m)}(p)$.

Definition: Under the hypotheses of Lemma 15.3, $f \in V_n$ is said to have a zero of order m in p if and only if $f^{[0]}(p) = \ldots = f^{[m-1]}(p) = 0 \neq f^{[m]}(p)$. An $f \in V$ is said to have m zeros counting multiplicities if and only if there exist $(p_1,\ldots,p_r) \in \Delta_r(M)$ such that f has a zero of order m_i in p_i, $i = 1,\ldots,r$, with $\sum_{i=1}^{r} m_i = m$.

Theorem 15.4: Let M be open and $f_o,\ldots,f_n \in C^k(M)$ an extended Markov system of order $k \leq n$. Then every $f \in V_n \setminus \{0\}$ has at most n zeros counting multiplicities.

Proof: For k = 1 all zeros counting multiplicities are of order 1 and coincide with usual zeros.
$k-1 \Rightarrow k$: If an $f \in V_n \setminus \{0\}$ had n+1 zeros counting multiplicities, $\left(f^{[0]}\right)'$ would have n zeros counting multiplicities, and so the same would hold for $f^{[1]} \not\equiv 0$ in contradiction to the induction hypothesis.

The results and definitions of this chapter are without difficulty carried over to half-open or closed intervals M. Formulations and proofs are left to the reader. (For theorem 15.1, for example, use the continuity of $\frac{f_i'}{f_2'}$ on M and apply lemma 8.7 a).)
We close this chapter by establishing a connection between the concept of extended Markov systems and determinants. We first recall and extend a notation from chapter 6:

Definition: For $M \subset \mathbb{R}$ and $p = 1,2,\ldots$, let

$$\overline{\Delta_p}(M) := \{(t_1,\ldots,t_p) \in M^p \mid t_1 \leq \ldots \leq t_p\},$$

and for $k \in \{0,1,\ldots,p-1\}$

$$\overline{\Delta_p}(M)^{(k)} = \{(t_1,\ldots,t_1,t_2,\ldots,t_2,\ldots,t_s,\ldots,t_s) \in \overline{\Delta_p}(M) \mid t_1<t_2<\ldots<t_s,$$

each t_i occurs r_i+1 times with $r_i \leq k$, $i = 1,\ldots,s$, and $\sum_{i=1}^{s}(r_i+1)=p\}$.

Theorem 15.5: Let $g_1,\ldots,g_n \in C^k(M)$ be a Čebyšev system. Then

$$\det{}^*\begin{pmatrix} g_1 \cdots\cdots\cdots\cdots\cdots\cdots\cdots\cdots g_n \\ t_1 \ldots t_1 \ldots t_2 \ldots t_2 \ldots t_s \ldots t_s \end{pmatrix} :=$$

$$= \begin{vmatrix} g_1(t_1) \ldots g_1^{(r_1)}(t_1) & g_1(t_2) \ldots g_1^{(r_2)}(t_2) \ldots g_1(t_s) \ldots g_1^{(r_s)}(t_s) \\ \vdots & \vdots \\ g_n(t_1) \ldots g_n^{(r_1)}(t_1) & g_n(t_2) \ldots g_n^{(r_2)}(t_s) \ldots g_n(t_s) \ldots g_n^{(r_s)}(t_s) \end{vmatrix}$$

has weakly constant sign on $\overline{\Delta_n}(M)^{(k)}$.

<u>Proof:</u> We shall show the assertion only for $r_1 = n-1$ since the argument will then be clear for all other cases. Let $(t_1,\ldots,t_n) \in \Delta_n(M)$ be fixed, and define

$$g(t) := \det\begin{pmatrix} g_1 \cdots\cdots g_n \\ t_1 \ldots t_{n-1}\, t \end{pmatrix} \quad \text{for } t \in M.$$

The mean-value theorem gives $g(t_n) = g(t_n)-g(t_{n-1})=g'(\xi_n)(t_n-t_{n-1})$ for a suitable $\xi_n \in (t_{n-1},t_n)$, and so

$$s := \operatorname{sign} g(t_n) = \operatorname{sign}\begin{vmatrix} g_1(t_1) \ldots g_1(t_{n-1}) & g_1'(\xi_n) \\ \vdots & \vdots \\ g_n(t_1) \ldots g_n(t_{n-1}) & g_n'(\xi_n) \end{vmatrix}.$$

Another application of the mean-value theorem yields

$$s = \operatorname{sign}\begin{vmatrix} g_1(t_1) \ldots g_1(t_{n-2}) & g_1'(\xi_{n-1}) & g_1'(\xi_n) \\ \vdots & & \\ g_n(t_1) \ldots g_n(t_{n-2}) & g_n'(\xi_{n-1}) & g_n'(\xi_n) \end{vmatrix}$$

for a suitable $\xi_{n-1} \in (t_{n-2},t_{n-1})$.

Repeating this procedure $\frac{n(n+1)}{2}$ times leads to

$$s = \operatorname{sign}\begin{vmatrix} g_1(t_1) & g_1'(\eta_2) \ldots g^{(n-1)}(\eta_n) \\ \vdots & \\ g_n(t_1) & g_n'(\eta_2) \ldots g_n^{(n-1)}(\eta_n) \end{vmatrix} \text{ for suitable } \eta_2,\ldots,\eta_n \in M \text{ with } t_1<\eta_2<\ldots<\eta_n<t_n.$$

If we now let t_n approach t_1 we see that $s \cdot \det{}^*\begin{pmatrix} g_1 \cdots\cdots g_n \\ t_1 \cdots\cdots t_1 \end{pmatrix} \geq 0$.

Theorem 15.6: Let $f_1,\dots,f_n \in C^k(M)$ be an extended Markov system of order k. Then

$$\det{}^*\begin{pmatrix} f_1 \dots\dots\dots\dots\dots\dots\dots\dots f_n \\ t_1\dots t_1 t_2 \dots t_2 \dots t_s \dots t_s \end{pmatrix} \text{ has strictly constant sign on } \overline{\Delta_n}(M)^{(k)}.$$

Proof: If the determinant would vanish for a $(t_1,\dots,t_1,t_2,\dots,t_2,\dots,t_s,\dots,t_s) \in \overline{\Delta_n}(M)^{(k)}$, there would be an $f \in \text{span}\{f_1,\dots,f_n\} \setminus \{0\}$ with n zeros counting multiplicities, i.e., $f(t_1) = \dots = f^{(r_1)}(t_1) = f(t_2) = \dots = f^{(r_2)}(t_2) = \dots = f(t_s) = \dots = f^{(r_s)}(t_s) = 0$ in contradiction to theorem 15.4.

Corollary 15.7: Let $f_1,\dots,f_n \in C^{n-1}(M)$ be an extended Markov system. Then

$$\det{}^*\begin{pmatrix} f_1 \dots f_n \\ t_1 \dots t_n \end{pmatrix} \text{ has strictly constant sign on } \overline{\Delta_n}(M).$$

Exercises: 1) Show that in the integral representation one may take pairwise distinct c's as the lower limits of the integrals.

2) Given the integral representation, under which conditions on the w_i's do the h_i's form a Markov system?

3) Prove lemma 15.3.

4) Show that theorem 15.5 holds for weak Čebyšev systems as well.

16. The Gauss Kernel transformation

Consider the functions $f_i \in C^\infty[0,\infty)$ defined by $f_i(x) = x^{\alpha_i}$ for $x \in [0,\infty)$, $i = 1,\dots,n$, where $\alpha_1,\dots,\alpha_n \in \mathbb{R}$ with $0 = \alpha_1 < \dots < \alpha_n$ are arbitrarily fixed.

Then $(D_+f)(x) = (D_-f)(x) = \frac{f'(x)}{f_2'(x)}$ exists for all $f \in \text{span}\{f_1,\dots,f_n\}$, and we have $(D_+f_i)(x) = x^{\alpha_i-\alpha_2}$ for $i = 2,\dots,n$. So induction via lemma 13.2 shows that $f_1,\dots,f_n$ is a normed Markov system, which, moreover, is extended. The same assertion clearly holds for $\alpha_1,\dots,\alpha_n \in \mathbb{R}$ with $\alpha_1 < \dots < \alpha_n$. So we get by a substitution $x = e^t$:

Lemma 16.1: Let $f_i \in C^\infty(\mathbb{R})$ be defined by $f_i(t) = e^{\alpha_i t}$, $i = 1,\dots,n$, where $(\alpha_1,\dots,\alpha_n) \in \Delta_n(\mathbb{R})$ is arbitrarily fixed. Then $f_1,\dots,f_n$ form an extended Markov system.

An elementary computation yields:

<u>Lemma 16.2:</u> Lemma 16.1 holds for $f_i(t) = e^{-(\alpha_i - t)^2}$, $i = 1,\dots,n$.

Clearly, t may be replaced by $\gamma^2 t$ for $\gamma \in (0,\infty)$. Hence follows:

<u>Corollary 16.3:</u> Lemma 16.1 holds for

$$f_i(t) = e^{-\gamma(\alpha_i - t)^2}, \quad i = 1,\dots,n,$$

where $\gamma \in (0,\infty)$ is arbitrarily fixed.

Especially we get from 15.7 that $\det^*\begin{pmatrix} f_1 \dots f_n \\ t_1 \dots t_n \end{pmatrix}$ has strictly constant sign on $\overline{\Delta_n(\mathbb{R})}$.

For the proof of the following result the reader is referred to POLYA and SZEGÖ (1972, vol. I, p. 61, problem 68) where a more general version is shown:

<u>Theorem 16.4:</u> Let M be a finite interval, $f_1,\dots,f_n \in C(M)$ bounded, $L : [a,b] \times [a,b] \to \mathbb{R}$ a continuous kernel and $g_1 \dots, g_n \in C(M)$ defined by

$$g_i(t) := \int_M f_i(s)\, L(s,t)\, ds.$$

Then

$$\det\begin{pmatrix} g_1 \dots g_n \\ t_1 \dots t_n \end{pmatrix} = \int_{(s_1,\dots,s_n) \in \Delta_n(M)} \cdots\cdots \int \det\begin{pmatrix} f_1 \dots f_n \\ s_1 \dots s_n \end{pmatrix} \cdot \det\begin{pmatrix} L(s_1,t_1) \dots L(s_1,t_n) \\ \vdots \quad\quad \vdots \\ L(s_n,t_1) \dots L(s_n,t_n) \end{pmatrix} ds_1 \dots ds_n$$

holds for all $(t_1,\dots,t_n) \in \Delta_n(\overline{M})$.

If we especially set $L(s,t) := e^{-\gamma(s-t)^2}$, we get from corollary 16.3:

<u>Theorem 16.5:</u> Let the hypotheses of theorem 16.4 be fulfilled, $f_1,\dots,f_n$ a weak Čebyšev system and $L(s,t) = \frac{k}{\sqrt{2\pi}} e^{-\frac{k^2}{2}(s-t)^2}$ with $k \in N$ fixed. Then $g_1,\dots,g_n \in C^\infty(M)$ form an extended Markov system.

If we consider L for varying k, i.e., $L_k(s,t) = \frac{k}{\sqrt{2\pi}} e^{-\frac{k^2}{2}(s-t)^2}$ and set

$$f_i^{(k)}(t) = \int_a^b f_i(s)\, L_k(s,t)\,ds \quad \text{for } i = 1,\dots,n \text{ and } k = 1,2,\dots,$$

it is not difficult to see that $f_i^{(k)}(t) \to f_i(t)$ holds for $k \to \infty$ and all $t \in M$.

If for every k there is a $g^{(k)}$ weakly adjoined to $U^{(k)} := \operatorname{span}\{f_1^{(k)},\dots,f_n^{(k)}\}$ and $g^{(k)}$ converges pointwise to a function $g \in C(M)$, g is weakly adjoined to $U := \operatorname{span}\{f_1,\dots,f_n\}$, for if for some $f \in U$, $g + f$ had a strong alternation of length $n + 2$, the same would hold for $g^{(k_o)} + f^{(k_o)}$ and k_o sufficiently large, where $f^{(k)} \in U^{(k)}$, $k = 1,2,\dots$, is a sequence converging to f.

17. Adjoined functions in the continuous case

As a preliminary result we show

Theorem 17.1: Let $f_1,\dots,f_n \in C^{n-1}(M)$ be an extended Markov system, $p \in \operatorname{int} M$ arbitrarily fixed and $f \in U_n \setminus \{0\}$ with a zero of order $n-1$ in p. Then the function $g \in C^{n-1}(M)$ defined by

$$g(t) = \begin{cases} 0 & \text{for } t < p \\ f(t) & \text{for } t \geq p \end{cases} \quad \text{for } n \geq 2 \quad \text{or} \quad g(t) = \begin{cases} 0 & \text{for } t < p \\ 1 & \text{for } t \geq p \end{cases} \quad \text{for } n=1$$

is weakly adjoined to U_n, and for every $(t_1,\dots,t_{n+1}) \in \Delta_{n+1}(M)$ with $t_n < p < t_{n+1}$, $g|_{\{t_1,\dots,t_{n+1}\}}$ is adjoined to $U|_{\{t_1,\dots,t_{n+1}\}}$.

Proof: We obviously need to consider only the case of an extended normed Markov system $f_1,\dots,f_n$. For $n = 1$ the statement is trivial.

$n-1 \Rightarrow n$: For every $t \in M$ we have $(Dg)(t) = \begin{cases} 0 & \text{for } t < p \\ (Df)(t) & \text{for } t \geq p. \end{cases}$

Df has a zero of degree $n-2$ in p, and $Df_2,\dots,Df_n \in C^{n-2}(M)$ form an extended normed Markov system. By induction hypothesis, Dg is weakly adjoined to $D\,U_n$. Lemma 13.2 shows that g is then weakly adjoined to U_n.

If for an $h \in U_n$, $h + \alpha g \not\equiv 0$ had zeros in $t_1,\dots,t_{n+1}$, $g(t_1) = \dots = g(t_n) = 0$ would yield $h(t_1) = \dots = h(t_n) = 0$, so $h \equiv 0$. But then $0 = g(t_{n+1}) = f(t_{n+1})$, and f would have n zeros counting multiplicities in contradiction to theorem 15.4.

Theorem 17.2: Let M be a bounded interval and $f_1,\dots,f_n \in C(M)$ a Markov system of bounded functions. Then there exists a bounded $r \in C(M)$ adjoined to $\operatorname{span}\{f_1,\dots,f_n\}$.

Proof: For n = 1 the assertion is trivial (use lemma 3.2). Now assume $n \geq 2$. Put a: = inf M, b: = sup M. The Gauss Kernel transformation yields extended Markov systems

$$f_1^{(k)},\dots,f_n^{(k)} \in C^\infty[a,b] \quad \text{for } k = 1,2,\dots \quad .$$

Let $p \in (a,b)$ be fixed, and define

$$g^{(k)}(t) = \begin{cases} 0 \text{ for } t \in [a,p) \\ f^{(k)}(t) \text{ for } t \in [p,b]. \end{cases}$$

where $f^{(k)} \in U_n^{(k)} \setminus \{0\}$ has a zero of order n-1 in p. By theorem 17.1 $\operatorname{span}\{U_n^{(k)}, g^{(k)}\}$ is a weak Haar space for each k = 1,2,... . Setting $U_n := \operatorname{span}\{f_1,\dots,f_n\}$ and $g := \lim_{k\to\infty} g^{(k)}$, the remarks after theorem 16.5 imply that $U_{n+1} := \operatorname{span}\{U_n, g\}$ is a weak Haar space.

For every $(t_1,\dots,t_{n+1}) \in \Delta_{n+1}(M)$ with $t_n<p<t_{n+1}$, $U_{n+1}\big|\{t_1,\dots,t_{n+1}\}$ is a Haar space, as can be shown in the following way:

Suppose $h + \alpha g \in U_{n+1} \setminus \{0\}$ vanishes on $\{t_1,\dots,t_{n+1}\}$, where $h \in U_n$. So we have $h(t_1) = \dots = h(t_n) = 0$ and $h \equiv 0$. We get $g(t_{n+1}) = 0$. Let $x \in (p,t_{n+1})$ be a point with $g(x) \neq 0$, say $g(x) > 0$ without loss of generality. For sufficiently large k, we would get $0 = g^{(k)}(p) < g^{(k)}(x) > g^{(k)}(t_{n+1})$, and $f^{(k)}$ would have a weak alternation of length n+1 in $p,\dots,p,x,t_{n+1}$, a contradiction.

Since $p \in (a,b)$ was chosen arbitrarily, we see that - after changing the sign of g, if necessary - for every $\theta := (t_1,\dots,t_{n+1}) \in \Delta_{n+1}(M)$ there is a $g_\theta \in C(M)$ with $||g_\theta||_\infty = 1$,

$$\det\begin{pmatrix} f_1 \dots\dots f_n g_\theta \\ t_1 \dots\dots t_{n+1}\end{pmatrix} > 0$$

and

$$\det\begin{pmatrix} f_1 \dots\dots f_n g_\theta \\ u_1 \dots\dots u_{n+1}\end{pmatrix} \geq 0 \text{ for all } (u_1,\dots,u_{n+1}) \in \Delta_{n+1}(M).$$

As g is continuous, the determinant is positive in a neighbourhood $U_\theta \subset \Delta_{n+1}(M)$ of θ. The family $\{U_\theta \mid \theta \in \Delta_{n+1}(M)\}$ is a cover of $\Delta_{n+1}(M)$ and contains a denumerable subcover $\{U_{\theta_\nu}\}_{\nu=1}^\infty$ of $\Delta_{n+1}(M)$. Setting

$$r(t) := \sum_{\nu=0}^{\infty} 2^{-\nu} g_{\theta_\nu}(t) \qquad \text{for } t \in M,$$

we get an $r \in C(M)$ with the desired properties.

Theorem 17.2 is readily generalized to

Theorem 17.3: Let M be an interval and $U \subset C(M)$ an n-dimensional Haar space. Then there exists an $f \in C(M)$ adjoined to U.

Proof: By an arctg change of variable the assertion is reduced to bounded M; let a: = inf M, b: = sup M. By theorem 7.7 $U|_{(a,b)}$ has a Markov basis $f_1|_{(a,b)},\dots,f_n|_{(a,b)}$. Division by $\sum_{i=1}^{n} |f_i|$ enables us to assume that U consists of bounded functions. The fact that $f_1,\dots,f_n$ perhaps is only a weak Markov system on M does not invalidate the arguments in the proof of theorem 17.2.

It is not known if theorem 17.3 may also be derived by the methods of chapter 14.

18. Adjoined functions in the periodic case

Bearing in mind that all periodic Haar spaces have odd dimension we have to modify the definition of adjoined functions accordingly:

Definition: Let $U \subset F_\tau$ be a (2n+1)-dimensional periodic Haar space. A pair f,g of functions in F_τ is called (weakly) adjoined to U if and only if span {U,f,g} is a (2n+3)-dimensional periodic Haar space.

We shall establish the existence of adjoined functions in this setting only for a fairly restricted class of periodic Haar spaces. There may be more general results (see the notes of this chapter).

The following result on non-periodic Haar spaces may be of some independent interest:

Lemma 18.1: Let $M \subset \mathbb{R}$ and $f_0, f_1 \in F$ be a normed Markov system. Then there exist bounded functions $u,v \in F$ such that f_0,u,v,f_1 is a normed Markov system; if M is an interval and $f_1 \in C(M)$, u,v may be chosen in C(M).

Proof: As f_1 is strictly monotone, a simple modification of lemma 14.2 shows that it is sufficient to consider the case $f_1(x) = x$ for all $x \in M$. Moreover, $M = \mathbb{R}$ may be assumed. Let $u,v \in C(\mathbb{R})$ be defined by

$$u(x) = \int_0^x e^{-t^2}\,dt,\quad v(x) = e^{-x^2} \quad \text{for } x \in \mathbb{R}.$$

By theorem 15.1 and lemma 13.2 the assertion is equivalent to showing that the functions $f_0, \frac{v'}{u'}, \frac{1}{u'}$ form a Markov system. We have $\frac{v'(x)}{u'(x)} = -2x$ and $\frac{1}{u'(x)} = e^{x^2}$ for all x. Again invoking theorem 15.1 and lemma 13.2, the assertion follows.

Lemma 18.2: Let [a,b] be a real interval, $f_0, f_1, \dots, f_n \in C^n[a,b]$ an extended Markov system on (a,b) and assume $f_i(a) = f_i(b) = 0$ for $i = 0,1,\dots,n-1$. Then there are $u,v \in C[a,b]$ vanishing in a and b such that $f_0, f_1, \dots, f_{n-1}, u, v, f_n$ is a Markov system on (a,b).

Proof: According to theorem 15.2, on (a,b) we form the functions $f_i^{[\nu]}$ for $i,\nu = 0,1,\dots,n$. By lemma 18.1 there exist $u^{[n-1]}, v^{[n-1]} \in C(a,b)$ such that $f_{n-1}^{[n-1]}, u^{[n-1]}, v^{[n-1]}, f_n^{[n-1]}$ is a normed Markov system and $\sup_{x \in (a,b)} \{|u^{[n-1]}(x)|, |v^{[n-1]}(x)|\} \le K$ for some suitable constant K.

Applying lemma 13.2 repeatedly, first with $g = f_{n-1}^{[n-2]}$, then with $g = f_{n-2}^{[n-3]}$ and so forth leads to a normed Markov system $f_o^{[o]}, f_1^{[o]}, \dots, f_{n-1}^{[o]}, u^{[o]}, v^{[o]}, f_n^{[o]}$ on (a,b) with

$\max\{|u^{[o]}(x)|, |v^{[o]}(x)|\} \le K \cdot |f_{n-1}^{[o]}(x)|$ for every $x \in (a,b)$. So the functions u, v defined by

$$\begin{pmatrix} u(x) \\ v(x) \end{pmatrix} = \begin{cases} \begin{pmatrix} 0 \\ 0 \end{pmatrix} & \text{for } x \in \{a,b\} \\ f_0(x) \cdot \begin{pmatrix} u^{[o]}(x) \\ v^{[o]}(x) \end{pmatrix} & \text{for } x \in (a,b) \end{cases}$$

have the desired properties.

Theorem 18.3: Let [a,b] be a real interval, $\tau := b-a$ and $U \subset C_\tau^{2n}(\mathbb{R})$ a (2n+1)-dimensional periodic Haar space such that no $h \in U \setminus \{0\}$ has more than 2n zeros on [a,b) counting multiplicities. Then there exist $u, v \in C_\tau(\mathbb{R})$ adjoined to U.

Proof: For n = 0, let $U := \operatorname{span}\{f_0\}$. Take $u(x) = f_0(x) \cos(\frac{2\pi x}{\tau})$, $v(x) = f_0(x) \cdot \sin(\frac{2\pi x}{\tau})$ for $x \in \mathbb{R}$.
Now assume $n \ge 1$. Let $V := \{f \in U | f(a) = 0\}$. By lemma 18.2 there exist two functions $u,v \in C_\tau(\mathbb{R})$ vanishing in a and b such that span{V,u,v} and span{U,u,v} are Haar spaces on (a,b) of dimensions

2n+2 and 2n+3 respectively. If an $h \in \mathrm{span}\{U,u,v\} \setminus \{0\}$ had 2n+3 zeros on $[a,b)$, one of these zeros would be a, so $h \in \mathrm{span}\{V,u,v\}$. But h would have 2n+2 zeros on (a,b), a contradiction.

19. Generalized convex functions

Though many results on generalized convex functions are implicitely contained in previous chapters it may be useful to restate them in a different setting.

As in chapter 15 we shall make the general hypothesis that M is an interval. Again, many of the concepts and results subsequently developed might be extended to more general sets M without difficulty.

The concept of convexity starts out from the following elementary definition:

An $f \in F$ is called convex if

$$(x_2-x_0)\ f(x_1) \le (x_2-x_1)\ f(x_0) + (x_1-x_0)\ f(x_2)$$

holds for all $(x_0,x_1,x_2) \in \Delta_3(M)$, and strictly convex if the inequality is strict on $\Delta_3(M)$.

If we denote by f_i, $i = 0,1,\dots,n$, the powers $f_i(x) = x^i$, $i = 0,1,\dots,n$, this definition may be extended to:

Definition: $f \in F$ is called convex (strictly convex) with respect to $f_0,f_1,\dots,f_n$ if the determinant

$$\det\begin{pmatrix} f_0 & f_1 & \dots & f_n & f \\ x_0 & x_1 & \dots & x_n & x_{n+1} \end{pmatrix}$$

is nonnegative (positive) for all $(x_0,\dots,x_{n+1}) \in \Delta_{n+2}(M)$. The set of all functions convex with respect to $f_0,f_1,\dots,f_n$ is denoted by $C(f_0,f_1,\dots,f_n)$.

This definition can without any difficulty be generalized to the case that $f_0,f_1,\dots,f_n$ form a Markov system on M (not necessarily consisting of continuous functions). So the terms "convex" and "strictly convex" with respect to a Markov system $f_0,f_1,\dots,f_n$ have - except, possibly, for a change of sign - the same meanings as the terms "weakly adjoined" and "adjoined" to $\mathrm{span}\{f_0,f_1,\dots,f_n\}$. The use of two different expressions for the same property reflects but historical developments.

(One may also consider Markov systems with the additional property that

$\det \begin{pmatrix} f_o & f_1 & \dots & f_i \\ x_o & x_1 & \dots & x_i \end{pmatrix}$ is positive for all $(x_o,\dots,x_i) \in \Delta_{i+1}(M)$ and $i = 0,1,\dots,n$. Following KARLIN and STUDDEN (1966), such Markov systems are usually called "complete Tchebycheff systems" or CT-systems.)

Lemma 19.1: a) $C(f_o,f_1,\dots,f_n)$ is a convex cone.
b) If M is compact and $f_o,f_1,\dots,f_n$ are continuous, the interior of $C(f_o,f_1,\dots,f_n)$ - with respect to the supremum norm on M - consists of all functions strictly convex with respect to $f_o,f_1,\dots,f_n$.

We first restate thorem 17.3:

Theorem 19.2: Let $M = [a,b]$ and $f_o,f_1,\dots,f_n \in C(M)$ be a Markov system. Then the interior of $C(f_o,f_1,\dots,f_n)$ is not empty.

From theorems 11.3 a) and 11.4 a) the following basic property of generalized convex functions is immediate:

Theorem 19.3: Let $M = (a,b)$, $f_o,f_1,\dots,f_n \in F$ be a normed Markov system and $f \in C(f_o,f_1,\dots,f_n)$. Then f has the same continuity properties as f_1, i.e., for every $x \in (a,b)$ we have that if f_1 is continuous from the right (left) at x, the same holds for f.

Corollary 19.4: Let $M = (a,b)$ and $f_o,f_1,\dots,f_n \in C(M)$ be a Markov system. Then $C(f_o,f_1,\dots,f_n)$ lies in $C(M)$.

Proof: Divide by f_1, apply theorem 19.3 and multiply by f_1.

In analogy to theorem 15.2, for $M = (a,b)$ and a Markov system $f_o,f_1,\dots,f_n$ let us define linear operators $D_+^{[i]}$, $D_-^{[i]}$, $i = 0,1,\dots,n$, by

$$\left(D^{[0]}_{\overset{+}{(-)}} g\right)(x) = \frac{g(x)}{f_o(x)} ,$$

$$\left(D^{[i]}_{\overset{+}{(-)}} g\right)(x) = \lim_{y \to x \overset{+}{(-)}} \frac{D^{[i-1]}_{\overset{+}{(-)}} g\,(y) - D^{[i-1]}_{\overset{+}{(-)}} g\,(x)}{D^{[i-1]}_{\overset{+}{(-)}} f_i\,(y) - D^{[i-1]}_{\overset{+}{(-)}} f_i(x)} , \quad i = 1,\dots,n,$$

wherever this is possible. Repeated application of theorems 11.3 and 11.4 yields:

Theorem 19.5: Let $M = (a,b)$, $f_0,f_1,\dots,f_n \in F$ be a Markov system and $f \in C(f_0,f_1,\dots,f_n)$.

a) Then $D_+^{[i]}f_i,\dots, D_+^{[i]}f_n$ form a normed Markov system for $i = 0,1,\dots,n$, and $D_+^{[i]}f \in C(D_+^{[i]}f_i,\dots,D_+^{[i]}f_n)$ for $i = 0,1,\dots,n$;

b) statement a) holds for "$D_-^{[i]}$" instead of "$D_+^{[i]}$", too.

Theorem 19.6: Let $M = (a,b)$, $f_0,f_1,\dots,f_n \in F$ be a Markov system, $1 \le k \le n-1$ fixed and assume $D_+^{[i-1]}f_i = D_-^{[i-1]}f_i \in C(M)$ for $i = 1,\dots,k$. Then we have:

a) $C(D_+^{[i]}f_i,\dots,D_+^{[i]}f_n) \subset C(M)$ for $i = 1,\dots,k-1$;

b) $C(D_+^{[k]}f_k,\dots,D_+^{[k]}f_n) \subset C_+(M)$, $C(D_-^{[k]}f_k,\dots,D_-^{[k]}f_n) \subset C_-(M)$;

c) the following statements (A), $(A_0),\dots,(A_k)$ are equivalent to each other:

(A): $f \in C(f_0,\dots,f_n)$

(A_i): $D_+^{[i]}f \in C(D_+^{[i]}f_i,\dots,D_+^{[i]}f_n)$, $i = 0,1,\dots,k$.

Proof: a) and b) follow directly from lemmas 12.1 and 12.2. c) is implied by theorem 19.5 and lemma 13.2.

Exercises: 1) Prove lemma 19.1.

2) In theorem 19.6, characterize those $f \in C(f_0,\dots,f_n)$ corresponding to $0 \in C(D_+^{[k]}f_k,\dots,D_+^{[k]}f_n)$. (Hint: Use lemma 13.2).

VI. Miscellaneous properties

20. Embedding a function into a Haar space

Recently there has been some interest in characterizing those functions $f \in F$ for which a Haar space $U \subset F$ exists with $f \in U$. From chapter 3 we know that if $U \subset F$ is an n-dimensional Haar space every $f \in U \setminus \{0\}$ has only weak alternations of finite length. We shall prove the following converse:

Theorem 20.1: Let $f \in F \setminus \{0\}$ be a function with a weak alternation of length n but with no weak alternation of length n+1.

a) Then there exists a Haar space containing f.

b) If f is bounded there is a Haar space of bounded functions containing f.

Proof: Let $Z(f) = \{z_1, \ldots, z_k\}$ with $z_1 < \ldots < z_k$ be the set of zeros of f. Clearly we have $0 \leq k \leq n-1$. We define an equivalence relation "$\sim$" on M by

$$x \sim y \Leftrightarrow x = y \in Z(f) \text{ or } f(x)\cdot f(t) > 0 \text{ for all } t \in \begin{cases} M \cap [x,y] & \text{if } x \leq y \\ M \cap [y,x] & \text{if } y \leq x \end{cases}$$

for all $x,y \in M$. From the hypothesis on f follows that in this way M is split into finitely many equivalence classes $M_1, \ldots, M_r$ such that for all $(x_1, \ldots, x_r) \in M_1 \times \ldots \times M_r$ we have $x_1 < \ldots < x_r$.
Let $s: M \to \mathbb{R}$ be defined by

$$s(x) = x + i \quad \text{for } x \in M_i,\ i = 1, \ldots, r \text{ (see figure 1).}$$

It is clearly sufficient to show the assertion for $\tilde{f} = f \circ s^{-1}: s(M) \to \mathbb{R}$. Whenever f has distinct nonzero sign on two consecutive sets $s(M_i)$ and $s(M_{i+1})$, $i = 1, \ldots, r-1$, define a point $p_i := \sup s(M_i) + \frac{1}{2}$ between them. Let q be a real polynomial of minimal degree d that has the same sign as $\tilde{f}$ on $s(M)$ and additional zeros in all the points p_i. We thus get $\tilde{f} = q \cdot h$ where $h = s(M) \to \mathbb{R}$ is defined by

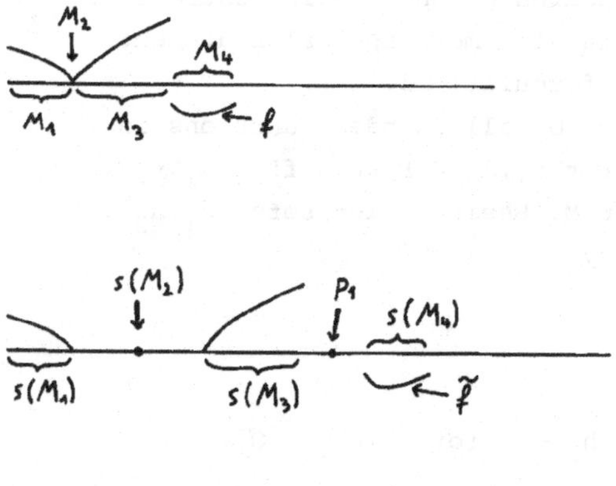

figure 1

$$h(x) = \begin{cases} 1 & \text{for } x \in s(M) \cap Z(q) \\ \frac{\tilde{f}(x)}{q(x)} & \text{for } x \in s(M) \setminus Z(q). \end{cases}$$

h is positive on s(M) and bounded if $\tilde{f}$ was bounded. As the real polynomials of degree d form a Haar space lemma 3.2 yields the assertion.

It is evident from the proof that the Haar space constructed has a Markov basis.
If $f \in F$ is a continuous function the construction used here will in general yield a Haar space of discontinuous functions. So the question remains open if an $f \in C(M)$ fulfilling the hypotheses of theorem 20.1 can be embedded into a Haar space $U \subset C(M)$.
Another interesting unsolved problem is the following: Let $n \geq 1$ be fixed and $V \subset F$ be a finite-dimensional linear space such that no $f \in V \setminus \{0\}$ has a weak alternation of length greater n. Does there always exist a Haar space U containing V?

Exercise: 1) Why does the construction in the proof of theorem 20.1 remain valid for unbounded sets M?
2) Show that in the proof of theorem 1 one in fact has $d = n-1$.
3) Give an example of functions $f,g \in C[a,b]$ each of which is embeddable into a Haar space but whose linear hull is not.

21. Oscillation theorems

In this chapter we shall deal with the possibility of constructing a function in a Haar space that "oscillates" between two given functions as often as possible in a sense that will be made precise subsequently. For reasons of presentation we shall - in contrast to most other parts of this book - begin with more special statements and proceed to increasingly general formulations.
We shall need some notations: The set of all bounded functions in F will be denoted by B(M). As usual, for $h_1,h_2 \in F$ we define $h_1 \underset{(=)}{\leq} h_2$ iff $h_1(t) \underset{(=)}{\leq} h_2(t)$ holds for all $t \in M$. Besides, let sets $[h_1,h_2]$, (h_1,h_2), $((h_1,h_2))$ in F be defined by

$$[h_1,h_2] := \{f \in F \mid h_1 \leq f \leq h_2\},$$

$$(h_1,h_2) := \{f \in F \mid h_1 < f < h_2\},$$

$$((h_1,h_2)) := \{f \in F \mid h_1 + \varepsilon < f < h_2 - \varepsilon \text{ for some } \varepsilon > 0\}.$$

Clearly we have $((h_1,h_2)) = \bigcup_{\varepsilon>0} (h_1+\varepsilon,h_2-\varepsilon) = \bigcup_{\varepsilon>0} [h_1+\varepsilon,h_2-\varepsilon]$.
Throughout this chapter we shall consider two functions $h_1,h_2 \in F$, arbitrarily fixed, with $h_1 < h_2$. To make some formulations shorter we shall also define $h_3,h_4,\ldots, \in F$ by $h_2 = h_4 = h_6 = \ldots$, $h_1 = h_3 = h_5 = \ldots$.
At various occasions we will make use of the following lemma:

<u>Lemma 21.1:</u> Let $h_1,h_2 \in F$ be fixed and $U \subset F$ an n-dimensional Haar space. Then $U \cap [h_1,h_2]$ is compact.

<u>Proof:</u> Let $N \subset M$ be a subset consisting of n points. Then $||f||: = \max_{n\in N} |f(x)|$ is a norm on U. Let $\{u_i\}_{i=1}^{\infty}$ be a sequence in $U \cap [h_1,h_2]$. We have $||u_i|| \leq \max_{x\in N} \{|h_1(x)|, |h_2(x)|\}$ for all i, so the sequence is bounded in norm. Considering a fixed basis of U it becomes evident that $\{u_i\}$ has a convergent subsequence; let us assume $u_i \to u \in U$ for $i \to \infty$ without loss of generality. As we have $h_1(t) \leq u_i(t) \leq h_2(t)$ for every $t \in M$ and norm convergence implies pointwise convergence, $h_1(t) \leq u(t) \leq h_2(t)$ holds for $t \in M$.

<u>Theorem 21.2:</u> Let $M = [a,b]$ be a closed interval, $h_1,h_2 \in C(M)$ and $U \subset C(M)$ an n-dimensional Haar space such that $U \cap [h_1,h_2]$ is not empty. Then the following statements hold:
a) There exists a unique $f \in U \cap [h_1,h_2]$ such that for some $(t_1,\ldots,t_n) \in \Delta_n(M)$ we have $f(t_i) = h_i(t_i)$ for $i = 1,\ldots,n$.
b) For the function f in a) there exists an additional point $t^* \in (-\infty,t_1) \cap M$ with $f(t^*) = h_2(t^*)$ or $t^* \in (t_n,\infty) \cap M$ with $f(t^*) = h_{n+1}(t^*)$ if and only if $U \cap (h_1,h_2)$ is empty.

<u>Proof:</u> a) By theorem 6.6 $U\setminus\{0\}$ contains a nonnegative element. So it is clear that the set

$$U_1: = \{f \in U \cap [h_1,h_2] | f(t) = h_1(t) \text{ for some } t \in M\}$$

is not empty. So the assertion is proved for n = 1. Now let n > 1, and define

$$r_1: = \inf \{t \in M | f(t) = h_1(t) \text{ for some } f \in U_1\}.$$

This infimum in fact is a minimum as can be seen immediately: Let $\{t_k\}_{k=1}^{\infty}$ be a decreasing sequence in M with $t_k \to r_1$ for $k \to \infty$, and

$\{f_k\}_{k=1}^{\infty}$ a sequence in U_1 with $f_k(t_k) = h_1(t_k)$ for $k = 1,2,\ldots$. As U_1 is compact by lemma 21.2, we may assume $f_k \to f \in U_1$ for $k \to \infty$ and a suitable $f \in U_1$. We have

$$|f(r_1) - h_1(r_1)| \le |f(r_1) - f(t_k)| + ||f-f_k|| + |f_k(t_k) - h_1(t_k)|$$

for all k, where $||\ ||$ is the supremum norm on M.
So the set $V_1: = \{f \in U_1 | f(r_1) = h_1(r_1)\}$ is not empty (and, trivially, compact). It is convenient to distinguish two cases:
1) If $a < r_1$, let $c: = a$ and $\varphi \in U \cap [h_0,h_1]$ be such that

$$\varphi(c) = \min\{\psi(c) | \psi \in U \cap [h_0,h_1]\}.$$

2) Let $a = r_1$: If the set $\{t \in (r_1,b] | f(t) = h_2(t)$ for some $f \in V_1\}$ is nonvoid, let r_2 be its infimum and $c \in (r_1,r_2)$ arbitrarily fixed. Otherwise let $c \in (a,b]$ be arbitrarily fixed. Let $\varphi \in V_1$ be such that $\varphi(c) = \max\{\psi(c) | \psi \in V_1\}$.
In both cases let k be maximal such that there exists $(t_1,\ldots,t_k) \in \Delta_k(M)$ with $\varphi(t_i) = h_i(t_i)$ for $i = 1,\ldots,k$. Suppose $k \le n-1$. For each $i \in \{1,\ldots,k-1\}$ define points s_i, u_{i+1} by

$$s_i = \max \{x \in [t_i,t_{i+1}) | \varphi(x) = h_i(x)\},$$

$$u_{i+1} = \min \{x \in (t_i,t_{i+1}] | \varphi(x) = h_{i+1}(x)\}.$$

We have $s_i < u_{i+1}$ for each i, since $s_i = u_{i+1}$ is excluded by $h_1 < h_2$ and $u_{i+1} < s_i$ for some i would yield a sequence of at least k + 2 touching points.
In case 1) we choose $z_0 \in (c,t)$ and $z_i \in (s_i,u_{i+1})$ for $i = 1,\ldots,k-1$. In case 2) we choose $z_0 = a$ and $z_i \in (s_i,u_{i+1})$ for $i = 1,\ldots,k-1$, with $c < z_1$, and $z_k = b$ if n-k is even. By theorem 6.6 a) and c) there exists a $g \in U$ with strong sign changes in all the z_i's, no other zeros and $g(t_1) > 0$.
So for sufficiently small $\varepsilon > 0$ we have

$$\varphi + \varepsilon g \begin{cases} \in U \cap (h_1,h_2) \text{ and } (\varphi + \varepsilon g)(c) < \varphi(c) \\ \in V_1 \qquad\qquad \text{ and } (\varphi + \varepsilon g)(c) > \varphi(c) \end{cases} \text{for case} \begin{cases} 1) \\ 2), \end{cases}$$

in either case a contradiction. So far, the existence part is proved.
The uniqueness part is easy to show:
If there were two functions $f,\hat{f} \in U \cap [h_1,h_2]$ and $(t_1,\ldots,t_n)$, $(\hat{t}_1,\ldots,\hat{t}_n) \in \Delta_n(M)$ with $f(t_i) = h_i(t_i)$, $\hat{f}(\hat{t}_i) = h_i(\hat{t}_i)$ for

$i = 1,\dots,n$ and $f \neq \hat{f}$, there would be a $j \in \{1,\dots,n\}$ with $t_i = \hat{t}_i$ for $i = 1,\dots,j-1$ and $t_j \neq \hat{t}_j$, say $t_j < \hat{t}_j$. Then $f - \hat{f}$ would have a weak alternation of length n+1 in $t_1,\dots,t_j,\hat{t}_j,\dots,\hat{t}_n$, a contradiction.

b) I) Assume $U \cap (h_1,h_2) \neq \emptyset$. If there were a point t^* with the described properties, $f - g \in U$ would have a strong alternation of length n+1 in $t^*,t_1,\dots,t_n$ or $t_1,\dots,t_n,t^*$ for every $g \in U \cap (h_1,h_2)$, a contradiction.

II) Assume $U \cap (h_1,h_2) = \emptyset$. For each $i \in \{1,\dots,n-1\}$ define points s_i,u_{i+1} as in the proof of part a). Let $z_i \in (s_i,u_{i+1})$ be arbitrarily fixed for $i = 1,\dots,n-1$ and $g \in U$ be defined by $g(z_1) = \dots = g(z_{n-1}) = 0$, $g(t_1) > 0$. If we had $f(x) < h_2(x)$ for all $x \in M \cap ((-\infty,t_1) \cup (t_n,\infty))$, $f + \varepsilon g$ would be contained in $U \cap (h_1,h_2)$ for sufficiently small $\varepsilon > 0$ in contradiction to the hypothesis.

If one wishes to generalize theorem 21.2 various limitations are ovserved as we shall illustrate by a few examples.

1) The hypothesis $h_1 < h_2$ cannot be weakend to $h_1 \leq h_2$: Let $M = [0,1]$, $-h_1(x) = h_2(x) = x^3$ and U the space of quadratic polynomials on M. We have $U \cap [h_1,h_2] = \{0\}$.

2) The hypothesis $h_1 < h_2$ implies $((h_1,h_2)) \neq \emptyset$ if M is compact and $h_1,h_2 \in C(M)$. In more general cases it is no longer sufficient to postulate $h_1 < h_2$ as the preceding example shows if M is replaced by $(0,1]$.

3) If M is no longer compact the functions in $U \subset C(M)$ need not be bounded. Let $M = [0,\frac{\pi}{2})$, $h_1(x) = 0$, $h_2(x) = 1$, $f(x) = \tan(x)$ for $x \in M$ and $U = \operatorname{span}\{h_2,f\}$.

The second example suggests a wider concept of touching points allowing for "asymptotic touching points" and also including boundary points of M. Such a concept is indispensable if we wish to deal with discontinuous functions. While various ideas might be used in this context the most general definition possible seems to be the following:

<u>Definition:</u> Let $h_1,h_2 \in F$ with $h_1 < h_2$ and $u \in [h_1,h_2]$. A point $(t_1,\dots,t_k) \in \overline{\Delta}_k(\overline{M})$ is called a lower t-oscillation of u relative to $[h_1,h_2]$ if for every $\varepsilon > 0$ there is $(p_1,\dots,p_k) \in \Delta_k(M)$ such that

$$|(u-h_i)(p_i)| < \varepsilon \text{ and } |p_i-t_i| < \varepsilon$$

holds for $i = 1,\dots,k$. k is called the length of the t-oscillation $(t_1,\dots,t_k)$.

$(t_1,\dots,t_k) \in \bar{\Delta}_k(\bar{M})$ is called an upper t-oscillation of u of length k relative to $[h_1,h_2]$ if $(t_1,\dots,t_k)$ is a lower oscillation of $-u$ relative to $[-h_2,-h_1]$.

This definition, however, turns out to be a bit too general for our purposes as the following example shows:

4) Let $M = (0,1]$, Q the space of quadratic polynomials on M, and

$$U = \{f \in C(M) \mid f(x) = (2+\cos(\tfrac{1}{x}))p(x) \text{ for some } p \in Q\}.$$

Let $h_1,h_2 \in C(M)$ be defined by $h_1(x) = 1 - x^3$, $h_2(x) = 3 + x^3$.
We have $U \cap [h_1,h_2] = \{f\}$ with $f(x) = 2 + \cos(\frac{1}{x})$ for $x \in M$, so no $f \in U \cap [h_1,h_2]$ has a lower t-oscillation of length 3 with respect to $[h_1,h_2]$.

In view of this example a classification of various types of discontinuities is useful:

<u>Definition:</u> Let $M \subset \mathbb{R}$ be a set, $f \in F$ and $t \in \bar{M}$. By $f_+\{t\}$, $f_-\{t\}$, $f_o\{t\}$, $f_{+,-}\{t\}$, $f_{+,o}\{t\}$, $f_{-,o}\{t\}$, $f\{t\}$ we denote the following <u>sets</u>:

$$f_+\{t\} = \{x \in \mathbb{R} \mid \text{there is a strictly decreasing sequence } \{t_k\}_{k=1}^{\infty} \text{ in } M \text{ with } \lim_{k\to\infty} t_k = t \text{ and } \lim_{k\to\infty} f(t_k) = x\},$$

$$f_-\{t\} = \{x \in \mathbb{R} \mid \text{there is a strictly increasing sequence } \{t_k\}_{k=1}^{\infty} \text{ in } M \text{ with } \lim_{k\to\infty} t_k = t \text{ and } \lim_{k\to\infty} f(t_k) = x\},$$

$$f_o\{t\} = \begin{cases} \{f(t)\} & \text{for } t \in M \\ \emptyset & \text{for } t \in \bar{M} \setminus M, \end{cases}$$

$$f_{+,-}\{t\} = f_+\{t\} \cup f_-\{t\}, f_{+,o}\{t\} = f_+\{t\} \cup f_o\{t\}, f_{-,o}\{t\} =$$

$$f_-\{t\} \cup f_o\{t\}, \ f\{t\} = f_{+,-}\{t\} \cup f_o\{t\}.$$

<u>Definition:</u> A set $U \subset F$ is said to fulfill condition (A) with respect to $[h_1,h_2]$ if for no $u \in U \cap [h_1,h_2]$ there is a $t \in \bar{M}$ with

$$0 \in (u-h_1)_-\{t\} \cap (u-h_2)_-\{t\} \text{ or } 0 \in (U-h_1)_+\{t\} \cap (u-h_2)_+\{t\}.$$

Lemma 21.3: Let $h_1,h_2 \in F$ and $U \subset F$ be an n-dimensional Haar space such that $U \cap ((h_1,h_2))$ is not empty. Then condition (A) is fulfilled.

Proof: Let $u \in U \cap [h_1,h_2]$ and $v \in U \cap [h_1+\varepsilon,h_2-\varepsilon]$ for some $\varepsilon > 0$. If (A) were violated, u - v would have a strong alternation of arbitrary length.

For the proof of the generalized version of theorem 21.2 we need some more auxiliary results:

Lemma 21.4: Let $u \in [h_1,h_2]$ be such that $\{u\}$ fulfills condition (A) and $(t_1,\dots,t_k) \in \overline{\Delta}_k(\overline{M})$ a lower t-oscillation of u relative to $[h_1,h_2]$. Then we have:

a) $t_1 \leq t_2 \leq \dots \leq t_k$ and $0 \in (u-h_j)\{t_j\}$ for $j = 1,\dots,k$.

b) If $t_{j-1} = t_j$ holds for some j, we have

$$0 \in (u-h_{j-1})_-\{t_j\} \cap (u-h_j)_+\{t_j\}, \text{ or}$$

$$0 = (u-h_{j-1})(t_j) \in (u-h_j)_+\{t_j\}, \text{ or}$$

$$0 = (u-h_j)(t_j) \in (u-h_{j-1})_-\{t_j\}.$$

c) If $t_{j-1} = t_j = t_{j+1}$ holds for some j, we have

$$0 = (u-h_j)(t_j) \in (u-h_{j-1})_-\{t_j\} \cap (u-h_{j-1})_+\{t_j\}$$

d) $t_j < t_{j+3}$ holds for all j.

Proof: b) By hypothesis there exist two sequences $\left\{p_{j-1}^{(n)}\right\}_{n=1}^{\infty}, \left\{p_j^{(n)}\right\}_{n=1}^{\infty}$ in M with

$$|(u-h_i)(p_i^{(n)})| < \frac{1}{n} \text{ for all n and } i = j-1,j,$$

$$p_{j-1}^{(n)} < p_j^{(n)} \text{ for all n,}$$

$$p_i^{(n)} \to t_i \text{ for } n \to \infty \text{ and } i = j-1,j.$$

Without loss of generality we can assume each of these sequences is strictly monotone or constant. By condition (A), $\left\{p_{j-1}^{(n)}\right\}$ and $\left\{p_j^{(n)}\right\}$ cannot be strictly increasing or strictly decreasing together. As $h_1 < h_2$ holds, only the three possibilities listed are left.

c) Considering b) for $t_{j-1} = t_j$ and $t_j = t_{j+1}$ and taking into account condition (A) one checks that the only possibility left is the one stated.

d) is immediate from c), a) is trivial.

Lemma 21.5: Let $t \in \bar{M}, h \in F$ be fixed, $U \subset B(M)$ a finite-dimensional linear space and $X \subset U$ compact. Then each of the sets

$$\{f \in X | 0 \in (f-h)_-\{t\}\}, \{f \in X | 0 \in (f-h)_+\{t\}\}, \{f \in X | 0 = (f-h)(t)\}$$

is compact.

Proof: Let $\left\{f_i\right\}_{i=1}^{\infty}$ be an sequence in X with $0 \in (f_i-h)_-\{t\}$ for all i. As X is compact we may without loss of generality assume $f_i \to f \in X$ for $i \to \infty$.

By hypothesis, for each i there is a strictly increasing sequence $\left\{t_k^{(i)}\right\}_{k=1}^{\infty}$ in M with $(f_i-h)(t_k^{(i)}) \to 0$ for $k \to \infty$. As the supremum norm is a norm on U, the inequality

$$|f(t_k^{(i)}) - h(t_k^{(i)})| \le ||f-f_i|| + |f_i(t_k^{(i)})|$$

shows that we can find a strictly increasing sequence $\left\{x_i\right\}_{i=1}^{\infty}$ in M with $x_i \to t$ for $i \to \infty$ and $|f(x_j)-h(x_j)| < \frac{1}{j}$ for all j. But this means $0 \in (f-h)_-\{t\}$.

The proof for the other sets is analogous.

The proof of the following lemma is left to the reader.

Lemma 21.6: Let $M \subset \mathbb{R}$ be bounded, $U \subset B(M)$ a finite-dimensional linear space and $X \subset U$ compact. Then the set $\{t \in \bar{M} | 0 \in f\{t\}$ for some $f \in X\}$ is compact.

Theorem 21.7: Let M be bounded and $h_1,h_2 \in F$ with $((h_1,h_2)) \neq \emptyset$. Let $U \subset B(M)$ be an n-dimensional Haar space such that $U \cap [h_1,h_2] \neq \emptyset$. Then the following statements hold:

a) If U fulfills condition (A) with respect to $[h_1,h_2]$ or if $U \cap ((h_1,h_2))$ is not empty, there exists an $f \in U \cap [h_1,h_2]$ with a lower t-oscillation of length n.

b) For the function f in a) there exists a lower or an upper t-oscillation of length n+1 only if $U \cap ((h_1,h_2))$ is empty.

Proof: a) By lemma 21.3, condition (A) is always fulfilled. As $U \setminus \{0\}$ contains a nonnegative element g, for any $v \in U \cap [h_1,h_2]$

there exists an $\alpha \in \mathbb{R}$ such that $(v-\alpha g) \in U \cap [h_1,h_2]$ and $0 \in (v-\alpha g-h_1)\{t\}$ for some $t \in \overline{M}$. So the set

$$V_o := \{f \in U \cap [h_1,h_2] \mid 0 \in (f-h_1)\{t\} \text{ for some } t \in \overline{M}\}$$

is not empty. Let $\{f\}_1^\infty$ be a sequence in V_o. By lemma 21.1 we may assume $f_k \to f \in U \cap [h_1,h_2]$ for $k \to \infty$. For each k let $x_k \in \overline{M}$ be such that $0 \in (f_k-h_1)\{x_k\}$. Since $\overline{M}$ is compact we may assume $x_k \to x \in \overline{M}$ for $k \to \infty$. For each k choose $y_k \in M$ with $|y_k-x_k| < \frac{1}{k}$ and $|(f_k-h_1)(y_k)| < \frac{1}{k}$. $|(f-h_1)(y_k)| \le ||f-f_k|| + |(f_k-h_1)(y_k)|$ holds for all k, and so we get $0 \in (f-h_1)\{x\}$. Thus V_o is compact. Define

$$r_1 := \inf\{t \in \overline{M} \mid 0 \in (f-h_1)\{t\} \text{ for some } f \in V_o\}.$$

By lemma 21.6 this infimum is a minimum. So the set $V_1 := \{f \in V_o \mid 0 \in (f-h_1)\{r_1\}\}$ is not empty and, by lemma 21.5, compact. For $i \ge 2$ we define

$$r_i := \inf\{t \in \overline{M} \cap (r_{i-1},\infty) \mid 0 \in (f-h_i)\{t\} \text{ for some } f \in V_{i-1}\},$$

$$V_i := \{f \in V_{i-1} \mid 0 \in (f-h_i)\{r_i\}\}, \text{ as long as } r_i < \infty.$$

It is left to the reader to show that the r_i's are minima of the corresponding sets, and that if V_l is not empty there is an $f \in V_l$ with a lower t-oscillation of length l in $r_1,\dots,r_l$.
Suppose no $f \in U \cap [h_1,h_2]$ has a lower oscillation of length n. So let $l < n$ be maximal with $V_l \neq \emptyset$. Define $r_o := -\infty$, $r_{l+1} := +\infty$, and let m be the smallest index such that $(-\infty,r_m) \cap M$ contains at least n+1 points (The case that M consists of n points would contradict our assumption because then an $f \in U \cap [h_1,h_2]$ with a lower t-oscillation of length n can obviously be found). With $p = n-m$, choose $c_1,\dots,c_p \in (-\infty,r_m) \cap M\setminus\{r_1,\dots,r_{m-1}\}$ such that $c_1<\dots<c_p$ and $M \cap (c_p,r_m) \neq \emptyset$. Define indices $i_o,i_1,\dots,i_m$ with $i_o = 0$, $i_m = p$ and

$$c_{i_j+1},\dots,c_{i_{j+1}} \in M \cap (r_j,r_{j+1}) \text{ for } j = 0,1,\dots,m-1,$$

and signs $\delta_1,\dots,\delta_p$ with $\delta_{i_j+1} = \dots = \delta_{i_{j+1}} = (-1)^{j+1}$ for $j = 0,1,\dots,m-1$. Now let $\varphi \in V_{m-1}$ be a function with

$$\beta := \sum_{i=1}^{p} (-1)^{\delta_i} \varphi(c_i) = \max\left\{ \sum_{i=1}^{p} (-1)^{\delta_i} \psi(c_i) \mid \psi \in V_{m-1}\right\}.$$

The existence of such a φ is guaranteed by the compactness of V_{m-1}.

Now let k be maximal such that φ has a lower t-oscillation $(t_1,\ldots,t_k) \in \bar{\Delta}_k(\bar{M})$ of length k. Clearly we have $t_i = r_i$ for $i = 1,\ldots,m-1$. For each $i \in \{1,\ldots,k-1\}$ define points s_i, u_{i+1} by

$$s_i = \sup\{x \in \bar{M} \cap [t_i,t_{i+1}) \,|\, 0 \in (\varphi-h_i)\{x\}\},$$

$$u_{i+1} = \inf\{x \in \bar{M} \cap (t_i,t_{i+1}] \,|\, 0 \in (\varphi-h_{i+1})\{x\}\}.$$

Suppose we had $u_{i+1} < s_i$ for some i. If we have $t_i = u_{i+1}$ the definition of u_{i+1} implies $0 \in (\varphi-h_{i+1})_+\{t_i\}$, so $0 \notin (\varphi-h_i)_+\{t_i\}$ by condition (A), and we get $0 \in (\varphi-h_i)_{-,0}\{t_i\}$. In an analogous way we get $0 \in (\varphi-h_i)_-\{t_{i+1}\} \cap (\varphi-h_{i+1})_{+,0}\{t_{i+1}\}$ if $s_i = t_{i+1}$. The cases $t_i < u_{i+1}$ and $s_i < t_{i+1}$ being trivial, we conclude that in any case φ has a lower t-oscillation of length k+2 in $t_1,\ldots,t_i,u_{i+1},s_i,t_{i+1},\ldots,t_k$, a contradiction. So we have $s_i \le u_{i+1}$ for all $i = 1,\ldots,k-1$.

Now let $z_0 \in (c_p,r_m) \cap M$ and $z_i \in [s_i,u_{i+1}] \cap \bar{M}$, $i = 1,\ldots,k-1$, be arbitrarily fixed. Setting $z_{-1} := -\infty$ and $z_k := \infty$, by theorem 6.7 there is a $g \in U\setminus\{0\}$ with $(-1)^j\, g(x) \ge 0$ for $x \in (z_{j-1},z_j) \cap M$, $j = 0,1,\ldots,k+1$. So we have $(-1)^{\delta_i}\, g(c_i) \ge 0$ for $i = 1,\ldots,p$ and $(-1)^{\delta_\nu}\, g(c_\nu) > 0$ for at least one index ν. For sufficiently small $\varepsilon > 0$ we thus have $\varphi + \varepsilon g \in V_{m-1}$ and

$$\sum_{i=1}^{p} (-1)^{\delta_i}\, (\varphi+\varepsilon g)(c_i) > \beta, \text{ a contradiction.}$$

b) The proof is the same as part I of the proof of theorem 21.2 b).

We believe theorem 21.7 b) can be sharpened to:

b') For some function f from a) there exists a lower or an upper t-oscillation of length n+1 if and only if $U \cap ((h_1,h_2))$ is empty.

However, we have not yet been able to prove this result.
The uniqueness of f in theorem 21.7 a) can no longer be guaranteed (it might still be true for $U \cap ((h_1,h_2)) \neq \emptyset$) as the following example shows:

<u>Example 21.8:</u> Let $M = \{0\} \cup (1,2]$, $h_1 \equiv 1$, $h_2 \equiv 2$ and $U = \text{span}\{f,g\}$

with

$$\binom{f}{g}(x) = \begin{cases} \binom{1}{0} & \text{for } x = 0 \\ \binom{2}{x-1} & \text{for } x \in (1,2]. \end{cases}$$

U is a Haar space, and we have $U \cap ((h_1,h_2)) = \emptyset$, $U \cap [h_1,h_2] = \{h \in U | h = f-\beta g \text{ for some } \beta \in [0,1]\}$. For $\beta \in [0,1)$, $f - \beta g$ has a lower t-oscillation of length 2; $f - g$ has a lower t-oscillation of length 3 relative to $[h_1,h_2]$.

So if we wish to retain the properties from theorem 21.2 we have to restrict the degree of generality. The following result will be exploited heavily in the next chapter:

Theorem 21.9: Let M be compact and $h_1,h_2 \in F$ with $((h_1,h_2)) \neq \emptyset$. Let $U \subset C(M)$ be an n-dimensional Haar space such that $U \cap [h_1,h_2] \neq \emptyset$.
a) There exists a unique $f \in U \cap [h_1,h_2]$ with a lower t-oscillation of length n.
b) The function f from a) has a lower or an upper t-oscillation of length n+1 if and only if $U \cap ((h_1,h_2))$ is empty.

Proof: a) From the hypotheses it is clear that for the lower t-oscillation $(t_1,\dots,t_n)$ of f we have $t_1<\dots<t_n$ and $f(t_i) \in h_i\{t_i\}$ for $i = 1,\dots,n$. Suppose there is an $\hat{f} \in U \cap [h_1,h_2]$ and a $(\hat{t}_1,\dots,\hat{t}_n) \in \Delta_n(M)$ with $\hat{f}(\hat{t}_i) \in h_i\{\hat{t}_i\}$ for $i = 1,\dots,n$ and $f \neq \hat{f}$; so there is a $j \in \{1,\dots,n\}$ with $t_j \neq \hat{t}_j$, say $t_j < \hat{t}_j$ without loss of generality. $t_1,\dots,t_j,\hat{t}_j,\dots,\hat{t}_n$ form a weak alternation of $f - \hat{f}$ of length n+1, because otherwise f or $\hat{f}$ would not be contained in $[h_1,h_2]$, and we arrive at a contradiction.
b) We only have to show the sufficiency part. So assume $U \cap (h_1,h_2) = \emptyset$. For each $i \in \{1,\dots,n-1\}$ define points s_i,u_{i+1} as in the proof of part a) of theorem 21.2. As f is continuous and $((h_1,h_2)) \neq \emptyset$, we have $s_i < u_{i+1}$ for all i. Let $g, \tilde{g} \in U$ be defined by $g(s_1) = \dots = g(s_{n-1}) = 0, \tilde{g}(u_2) = \dots = \tilde{g}(u_n) = 0$, $g(t_1) = \tilde{g}(t_1)=1$. From the continuity of the functions in U and the compactness of M it follows that $h:= g + \tilde{g}$ is bounded away from zero on each of the sets $M_i:= M \cap [s_i,u_i]$, $i = 1,\dots,n$, positive on $M_1 \cup M_3 \cup \dots$ and negative on $M_2 \cup M_4 \cup \dots$. It is not difficult to see that $f + \varepsilon h$ is contained in $U \cap ((h_1,h_2))$ for sufficiently small $\varepsilon > 0$, a contradiction.

<u>Exercise:</u> Show theorem 21.9 without recurring to theorem 21.7.

22. Applications to uniform approximation

The results of the last chapter have some immediate consequences in the theory of uniform approximation. If $F' \subset F$ is a set endowed with some norm $\| \ \|$ and $U \subset F$ an arbitrary nonempty subset, an element $u^* \in U$ is called a best approximation or proximum to a $g \in F$ from U with respect to this norm if

a) $u - g \in F'$ for all $u \in U$, and

b) $\|u^*-g\| = \inf\{\|u-g\| \mid u \in U\}$.

In the case that U is a finite-dimensional linear space in F' it is well known that there exists a proximum to every $g \in F$ from U if a) holds. The existence of a best approximation to a given $g \in F$ does not constitute a problem in any of the situations to be considered subsequently.

We shall only deal with compact sets $M \subset \mathbb{R}$ and spaces $U \subset C(M)$, though some generalizations might clearly be given.

By $\| \ \|$ we denote the supremum norm, i.e., $\|f\| := \sup\{|f(x)| \mid x \in M\}$ for $f \in B(M)$.

<u>Theorem 22.1 (ordinary uniform approximation):</u> Let $M \subset \mathbb{R}$ be compact, $U \subset C(M)$ an n-dimensional Haar space and $g \in C(M)$ arbitrarily fixed. Then there exists exactly one best approximation $u \in U$ to g with respect to the supremum norm, and u is characterized by the property that there exists $(t_1,\dots,t_{n+1}) \in \Delta_{n+1}(M)$ with

$$-\|u-g\|^2 = (u-g)(t_i)\cdot(u-g)(t_{i+1}) \quad \text{for } i = 1,\dots,n.$$

<u>Proof:</u> Let u be a best approximation to g. If $u = g$, nothing has to be proved. For $u \neq g$, let $\lambda := \|u-g\| > 0$. So we have $u \in U \cap [g-\lambda, g+\lambda]$ and $U \cap (g-\lambda, g+\lambda) = \emptyset$. Now set $h_1 := g-\lambda$, $h_2 := g+\lambda$ and apply theorem 21.9.

<u>Theorem 22.2 (one-sided uniform approximation):</u> Under the hypotheses of theorem 22.1 there exists exactly one $u \in U$ with $u \geq g$ and

$$\|u-g\| = \inf\{\|v-g\| \mid v \in U \text{ and } v \geq g\}.$$

u is characterized by the property that there is a $(t_1,\dots,t_{n+1}) \in \Delta_{n+1}(M)$ with

$$(u-g)(t_i) = \begin{cases} \| u^*-g \| & \text{for } i \text{ even} \\ 0 & \text{for } i \text{ odd, } i = 1,\dots,n+1 \end{cases}$$

or

$$(u-g)(t_i) = \begin{cases} 0 & \text{for } i \text{ even} \\ \| u^*-g \| & \text{for } i \text{ odd, } i = 1,\dots,n+1. \end{cases}$$

Proof: The proof is analogous to the proof of theorem 22.1, with $\lambda := \| u-g \|$, $h_1 := g$, $h_2 := g + \lambda$.

Theorem 22.3 (weighted uniform approximation): Let the hypotheses of theorem 22.1 be fulfilled and $w \in C(M)$ a positive function. Let $\| \; \|_w$ be defined by $\| f \|_w := \sup\{w(x)|f(x)| \,|\, x \in M\}$ for $f \in B(M)$. Then there is exactly one $u \in U$ with

$$- \| u-g \|_w = \inf\{ \| v-g \|_w \,|\, v \in U\},$$

and u is characterized by the property that there is a $(t_1,\dots,t_{n+1}) \in \Delta_{n+1}(M)$ with

$$\| u-g \|_w^2 = (u-g)(t_i)\cdot w(t_i)\cdot(u-g)(t_{i+1})\cdot w(t_{i+1}) \text{ for } i = 1,\dots,n.$$

Proof: Let $U_w := \{f \in C(M) | f = w \cdot h \text{ for an } h \in U\}$, and $g_w = g \cdot w$. U_w is an n-dimensional Haar space (lemma 3.2). Apply theorem 22.1 to U_w and g_w.

Theorem 22.4 (one-sided weighted approximation): Formulation and proof of this result are left to the reader.

Theorem 22.5 (simultaneous approximation): Let $M \subset \mathbb{R}$ be compact and $U \subset C(M)$ an n-dimensional Haar space. Let two families $h_i^\alpha = h_i(\cdot,\alpha) : M \times \mathbb{R} \to \mathbb{R}$, $i = 1,2$, of functions continuous in both arguments be given with the following properties:

a) There exists a $\beta^* \in \mathbb{R}$ such that $h_1^\alpha \le h_1^\beta < h_2^\beta \le h_2^\alpha$ holds for all $\alpha,\beta \in \mathbb{R}$ with $\beta^* \le \beta < \alpha$, and $U \cap [h_1^{\beta^*},h_2^{\beta^*}] = \emptyset$.

b) $U \cap [h_1^\alpha,h_2^\alpha]$ is not empty for α sufficiently large.

Then there exists a minimal $\gamma \in \mathbb{R}$ such that $A := U \cap [h_1^\gamma,h_2^\gamma]$ is not empty, A consists of exactly one element u, and u is characterized by the property that there is a $(t_1,\dots,t_{n+1}) \in \Delta_{n+1}(M)$ forming a lower or upper t-oscillation of u relative to $[h_1^\gamma,h_2^\gamma]$.

Proof: Let $C: = \{\alpha \in [\beta^*,\infty) | U \cap [h_1^\alpha, h_2^\alpha] \neq \emptyset\}$. One easily checks that C is compact. Let $\gamma: = \min(C)$. We have $U \cap (h_1^\gamma, h_2^\gamma) = \emptyset$ and can apply theorem 21.9.

Corollary 22.6: Let M be compact, $U \subset C(M)$ an n-dimensional Haar space and $f_-, f_+ \in C(M)$ with $f_- < f_+$ arbitrarily fixed. Define $s: U \to (0,\infty)$ by

$$s(v): = \max\{\| f_- - v\|, \| f_+ - v\|\} \quad \text{for } v \in U,$$

and let $u \in U$ be such that $s(u) = \inf\{s(v) | v \in U\}$. Then we either have $s(u) = \frac{1}{2}\| f_+ - f_-\|$, or u is unique and is characterized by the property that there is a $(t_1,\dots,t_{n+1}) \in \Delta_{n+1}(M)$ forming a lower or upper t-oscillation of u relative to $[f^+ - s(u), f^- + s(u)]$.

Proof: We have $s(u) \geq \beta^*: = \frac{1}{2}\|f_+ - f_-\|$. Let us define $h_1^\lambda: = f_+ - \lambda$, $h_2^\lambda: = f_- + \lambda$ for $\lambda \in \mathbb{R}$. Now either the hypotheses of theorem 22.5 are fulfilled or we have $U \cap [f_+ - \beta^*, f_- + \beta^*] \neq \emptyset$, implying $s(u) = \beta^*$ for every $u \in U \cap [f_+ - \beta^*, f_- + \beta^*]$.

For the rest of this chapter we return to ordinary uniform approximation. It is interesting to compare theorem 22.1 with some related results about weak Haar spaces. A best approximation $u^* \in U$ to some $g \in F$ will be said to have the "extremal alternation" property (EA) iff

$$-\| u^* - g\|^2 = (u^*-g)(t_i) \cdot (u^*-g)(t_{i+1}) \quad \text{for } i = 1,\dots,n$$

holds for a $(t_1,\dots,t_{n+1}) \in \Delta_{n+1}(M)$. The set of all proxima from U to some $g \in F$ will be denoted by $P_U(g)$.
JONES and KARLOVITZ (1970) proved the following

Theorem 22.7: Let M be a closed interval and $U \subset C(M)$ an n-dimensional linear space. Then the following properties are equivalent:
a) U is a weak Haar space.
b) For every $g \in C(M)$ there is a $u^* \in P_U(g)$ with property (EA).

The implication a) => b) of this result was generalized by NÜRNBERGER (1978) and DEUTSCH, NÜRNBERGER and SINGER (1978) to arbitrary compact sets M. Moreover, NÜRNBERGER showed:

Theorem 22.8: Let M be compact and $U \subset C(M)$ an n-dimensional weak Haar space. Then the following properties are equivalent:

a) No $f \in U \setminus \{0\}$ has more than n zeros.

b) For every $g \in C(M)$ there is exactly one $u^* \in P_U(g)$ with property (EA).

Combining theorem 22.1 with Haar's theorem (see, e.g. CHENEY 1966, p. 81) we get:

Theorem 22.9: Let M be compact and $U \subset C(M)$ an n-dimensional weak Haar space. Then the following properties are equivalent:

a) No $f \in U \setminus \{0\}$ has more than n-1 zeros (i.e., U is a Haar space).

b) For every $g \in C(M)$ the set $P_U(g)$ consists of one element.

Moreover, we have:

Theorem 22.10: Let M be compact and $U \subset C(M)$ an n-dimensional weak Haar space. Then for no $g \in C(M)$ there exists a $u \in U \setminus P_U(g)$ with property (EA).

Proof: Let $u^* \in P_U(g)$, $u \in U \setminus P_U(g)$ and $\lambda := \|u^*-g\|$. So we have $\tilde{\lambda} := \|u-g\| > \lambda$, $u \in U \cap [g-\tilde{\lambda}, g+\tilde{\lambda}]$ and $u^* \in U \cap (g-\tilde{\lambda}, g+\tilde{\lambda})$. If u had property (EA), $u - u^* \in U$ would have a strong alternation of length n+1 in the corresponding points, a contradiction.

As theorem 22.1 and related results could be derived immediately from the oscillation properties of chapter 21, we would like to make the following

Conjecture 22.11: Let M be compact, $U \subset C(M)$ an n-dimensional weak Haar space and $h_1, h_2 \in C(M)$ with $U \cap [h_1,h_2] \neq \emptyset$. Then the following assertions hold:

a) There exists an $f \in U \cap [h_1,h_2]$ with a lower oscillation of length n.

b) The set $U \cap (h_1,h_2)$ is empty if and only if there exists an $f \in U \cap [h_1,h_2]$ with a lower or upper t-oscillation of length n+1.

c) The function $f \in U \cap [h_1,h_2]$ from b) is unique if and only if no $f \in U \setminus \{0\}$ has more than n zeros.

Part c) of the conjecture would yield theorem 22.8, part b) would imply theorem 22.7, a) => b).

Weak Haar spaces of dimension n in which no nonvanishing function has more than n zeros, have received some attention in connection with continuous selections of the metric projection (see, e.g., NÜRNBERGER

1977). The relation between these spaces and Haar spaces is fairly close, as the following application of some results of chapter 4 shows:

Lemma 22.12: Let $U \subset F$ be an n-dimensional weak Haar space, but not a Haar space, and assume no $f \in U \setminus \{0\}$ has more than n zeros. Then the following statements hold:

a) If M contains an unessential point p (with respect to U), $U|_{M\setminus\{p\}}$ is a Haar space.

b) If there is an $f \in U$ with n separated zeros, M contains its infimum a and its supremum b, and we have $f(a) = f(b) = 0$. $U|_{M\setminus\{a\}}$ and $U|_{M\setminus\{b\}}$ are Haar spaces.

c) If M has property (B), there is an $f \in U$ with n separated zeros.

Exercises: 1) Find an n-dimensional weak Haar space U in which no $f \in U\setminus\{0\}$ has more than n zeros but there exists an $f \in U\setminus\{0\}$ with a weak alternation of length n+2.

2) Find a weak Haar space U with an $f \in U$ that changes sign strongly in an unessential point.

3) Find a weak Haar space and an essential point x such that no $f \in U$ changes sign strongly in x.

23. Disconjugacy

In this chapter we shall deal with some questions originating in the theory of ordinary differential equations.
Throughout this chapter M will always denote an interval and $U \subset C^n(M)$ an n-dimensional linear space, $n \geq 1$.

Definition: U is called a disconjugate space if and only if every $f \in U \setminus \{0\}$ has at most n-1 zeros counting multiplicities. A basis $f_1,\dots,f_n$ of a disconjugate space U is called a T^*-system. $f_1,\dots,f_n$ are called an M^*-system if $f_1,\dots,f_i$ is a T^*-system for $i = 1,\dots,n$.

Clearly a disconjugate space is a Haar space. The connection to differential equations is established by the following

Lemma 23.1: Let U be a disconjugate space. Then there exist $p_1,\dots,p_n \in C(M)$ and an n-th order linear differential operator $L: C^n(M) \to C(M)$ defined by

$$Ly: = y^{(n)} + p_1 \cdot y^{(n-1)} + \dots + p_{n-1}y' + p_n y \quad \text{for } y \in C^n(M) \qquad (*)$$

with $U = \ker(L)$.

Proof: Let $f_1,\dots,f_n$ be a basis of U and

$$W(t): = \det \begin{pmatrix} f_1(t) & \dots & f_n(t) \\ f_1'(t) & & f_n'(t) \\ \vdots & & \vdots \\ f_1^{(n-1)}(t) & \dots & f_n^{(n-1)}(t) \end{pmatrix} \quad \text{for } t \in M$$

the corresponding Wronskian. Define L by

$$(Ly)(t) = \frac{1}{W(t)} \det \begin{pmatrix} f_1(t) & \dots\dots & f_n(t) & y(t) \\ f_1'(t) & \dots\dots & f_n'(t) & y'(t) \\ \vdots & & \vdots & \vdots \\ f_1^{(n)}(t) & \dots\dots & f_n^{(n)}(t) & y^{(n)}(t) \end{pmatrix}$$

for $t \in M$ and $y \in C^n(M)$. One easily checks that L has the desired properties.

Definition: An (n-th order linear) differential operator L of the form (*) is called disconjugate if and only if ker (L) is a disconjugate space.

The assertion of lemma 23.1 becomes invalid if U is only assumed to be a Haar space. Take, for example, $M = \mathbb{R}$ and $U = \text{span}\{f,g\}$ with $f(t) \equiv 1$, $g(t) = t^3$.

<u>Theorem 23.2:</u> Let M be open or half-open and $U = \ker(L)$ for a differential operator L of the form (*), and $U|_{\text{int}(M)}$ a Haar space. Then U is a disconjugate space.

For a proof, the reader is kindly referred to COPPEL (1971; p. 82, proposition 3, and p. 102, theorem 8).

<u>Theorem 23.3:</u> Let M be open and U a disconjugate space. Then U has an M^*-basis.

<u>Proof:</u> Let $q, p \in C^{(n)}(M)$ be chosen with $p > 0$. Then q has exactly k zeros counting multiplicities if and only if $\frac{q}{p}$ has exactly k zeros counting multiplicities.
By theorem 7.7 U has a Markov basis $f_1,\ldots,f_n$. By the preceding argument we may without loss of generality assume $f_1 \equiv 1$. For <u>n = 1</u> the assertion is trivial.
<u>n-1 => n:</u> First suppose $f_2'(x) = 0$ for some $x \in M$. Let $f \in U \setminus \{0\}$ be a function with n-1 zeros, one of these being x. By theorem 11.3 there exists $(D_+f)(x) = \lim\limits_{t \to x+} \frac{f(t) - f(x)}{f_2(t) - f_2(x)}$, and we get $f'(x) = 0$. So f has n zeros counting multiplicities, a contradiction.
So we have $f_2' > 0$ without loss of generality and $D_+f = \frac{f'}{f_2'}$ for $f \in U$. D_+U is an (n-1)-dimensional Haar space in $C^{n-1}(M)$. By theorem 23.2 $U' := \text{span}\{f_2',\ldots,f_n'\}$ is disconjugate because it is a Haar space and

$$\det\begin{pmatrix} f_2'(t) & \cdots & f_n'(t) \\ \vdots & & \vdots \\ f_2^{(n-1)}(t) & \cdots & f_n^{(n-1)}(t) \end{pmatrix} = \det\begin{pmatrix} f_1(t) & \cdots & f_n(t) \\ \vdots & & \vdots \\ f_1^{(n-1)}(t) & \cdots & f_n^{(n-1)}(t) \end{pmatrix} = W(t).$$

So by the first paragraph of this proof D_+U is disconjugate, too. By induction hypothesis D_+U has an M^*-basis $\frac{g_2'}{f_2'},\ldots,\frac{g_n'}{f_2'}$, so each of the spaces $U_i' := \text{span}\{g_2',\ldots,g_i'\}$, $i = 2,\ldots,n$, is disconjugate.
$f_1, g_2,\ldots,g_n$ is a Markov system because of lemma 13.2, and

$$\det\begin{pmatrix} f_1(t) & g_2(t)\ldots\ldots g_i(t) \\ \vdots & \vdots \qquad \vdots \\ f_1^{(i-1)}(t) & g_2^{(i-1)}(t)\ldots g_i^{(i-1)}(t)\end{pmatrix} = \det\begin{pmatrix} g_2'(t)\ldots\ldots g_i'(t) \\ \vdots \qquad \vdots \\ g_i^{(i-1)}(t)\ldots g_i^{(i-1)}(t)\end{pmatrix} \neq 0$$

for $i = 2,\ldots,n$ implies via theorem 23.2 that $f_1, g_2, \ldots, g_n$ is an M^*-system.

Theorem 23.4: Let $M = [a,b]$ and $\tilde{M} = [c,d] \subset (a,b)$ with $c \leq d$. If $U|_{\tilde{M}}$ is disconjugate there is an $\varepsilon > 0$ such that $U|_{(c-\varepsilon,d+\varepsilon)}$ is disconjugate.

Proof: Let $f_1,\ldots,f_n$ be a basis of U, and define W and L as in the proof of lemma 23.1 where this is possible. $W: M \to \mathbb{R}$ is continuous and bounded away from zero on $\tilde{M}$, say, positive. So W is positive on $\hat{M} := [c-\varepsilon, d+\varepsilon]$ for some $\varepsilon > 0$. By Rolle's theorem $\hat{U} := U|_{\hat{M}}$ then is a Haar space, and we have $\hat{U} = \ker(L)$. Now apply theorem 23.2.

Corollary 23.5: Under the hypotheses of theorem 23.4, $U|_{\tilde{M}}$ has an M^*-basis.

Proof: Take $p \in [c-\varepsilon, c)$ and define

$$U_i := \{f \in U \mid f(p) = f'(p) = \ldots = f^{(n-1-i)}(p) = 0\}$$

$$\text{for } i = 1,\ldots,n-1.$$

Theorem 23.6: Let M be closed and U a disconjugate space. Then U has an M^*-basis.

Proof: Extend the domain of definition of U to $\mathbb{R}$ by setting $f(x) = f(a)$ for $x < a$ and $f(x) = f(b)$ for $x > b$. Apply corollary 23.5.

In contrast, examples 10.1 and 10.2 show that if M is half-open, for every $k \geq 2$ there is a disconjugate space of dimension k without a Markov basis.

Definition: Two points $c, d \in M$ are called conjugate points of U if $U|_{[c,d)}$ is disconjugate but $U|_{[c,d]}$ is not.

Theorem 23.7: Let $c, d \in M$ be conjugate points of U. Then the following properties are equivalent:

a) $\dim U|_{\{c,d\}} = 2$.

b) $U|_{[c,d]}$ is a Haar space.

Proof: a) => b) is immediate from theorem 4.8 e), b) => a) from lemma 1.2 c).

If $c, d \in M$ are conjugate points of U it is evident from theorem 23.4 that there is no larger interval $\tilde{M}$, i.e. $[c,d] \underset{\neq}{\subset} \tilde{M}$, such that U is a Haar space on $\tilde{M}$.

If a differential operator L of the form (*) is given, it would be interesting to determine its maximal range of disconjucagy or at least have lower bounds. However, the estimates known are rather poor (see COPPEL 1971, p. 84 - 86).

Exercises: We set $M = [0,\infty)$, $c = 0$ and $U = \ker(L) = \operatorname{span}\{f_1,\dots,f_n\}$.

1) Let $Ly = y^{(2)} - y$. Then $f_1(x) = \sin(x)$, $f_2(x) = \cos(x)$ and $d = \pi$. $U|_{[0,\pi)}$ has no M^*-basis. $U|_{[0,\pi]}$ is not a Haar space.

2) Let $Ly = y^{(3)} - y'$. Then $f_1(x) \equiv 1$, $f_2(x) = \sin(x)$, $f_3(x) = \cos(x)$ and $d = 2\pi$.

a) $U|_{[0,2\pi)}$ has no M^*-basis, $U|_{[0,2\pi]}$ is not a Haar space.

b) $U|_{(0,2\pi)}$ has a Markov basis $f_1 - f_3$, f_2, f_1. The space $\tilde{U} = \operatorname{span}\{f_1 - f_3, f_2\}$ is generated by $\tilde{L}$ with

$$(\tilde{L}y)(x) = y^{(2)}(x) + \frac{\sin(x)}{1-\cos(x)}\, y'(x) + \frac{1}{1-\cos(x)}\, y(x).$$

3) Let $Ly = y^{(4)} - y^{(2)}$. Then $f_1(x) \equiv 1$, $f_2(x) = x$, $f_3(x) = \sin(x)$, $f_n(x) = \cos(x)$ and $d = 2\pi$.

a)* $U|_{[0,2\pi]}$ is a Haar space with a Markov basis. Find such a basis.

b)* Show that $U|_{[0,2\pi]}$ may be extended to a Haar space on a larger domain $[0,2\pi] \cup \{p\}$, $p \notin [0,2\pi]$ by setting

$$\begin{pmatrix} f_1 \\ f_2 \\ f_3 \\ f_4 \end{pmatrix}(p) = \begin{pmatrix} 0 \\ 1 \\ 1 \\ 0 \end{pmatrix} .$$

c)* Show that $U|_{[0,2\pi]}$ cannot be extended to a Haar space on a domain augmented by two points (NEMETH 1974).

4) Let $L = y^{(5)} - y^{(3)}$. Then $f_i(x) = x^{i-1}$ for $i = 1,2,3$, $f_4(x) = \sin(x)$, $f_5(x) = \cos(x)$.

a) Find a Markov basis of $U\big|_{[0,2\pi]}$.

b)* Show that $2\pi < d$. (It is remarkable that even in this simple case the exact value of d is not known).

Notes

Chapter 1: Most of the material is found in classical textbooks on approximation theorey, the setting usually is more special.
Theorem 1.7 is found in STOCKENBERG (1976), the second proof is ours.

Chapter 2: Topological properties of sets admitting generalized Haar spaces were first investigated by MAIRHUBER (1956) who proved theorem 2.3 b) for the case $M \subset \mathbb{R}^n$. Independently this result was found by SIEKLUCKI (1958) and CURTIS (1959) for arbitrary compact Hausdorff spaces M. Simpler proofs were given by SCHOENBERG and YANG (1961) and ZIELKE (1973). The generalization to arbitrary Hausdorff spaces M is due to LUTTS (1964). Our proof is a simplification of an argument given by STOCKENBERG (1976) which - in turn - is based on LUTTS (1964) and ZIELKE (1973).
The definition of generalized Haar spaces can readily be carried over to the case of complex-valued functions. All results from chapter 1, lemmas 2.1 and 2.2 remain valid. It seems plausible to expect a Mairhuber-type result in this situation, too. First results were given by SCHOENBERG and YANG (1961). Their work was continued by OVERDECK (1971). The most general result to date is due to HENDERSON and UMMEL (1973):
Let M be a locally compact topological space. On M there exists an n-dimensional generalized Haar space of continuous complex-valued functions with $n \geq 2$ if and only if M is homeomorphic to a subset of the complex plane.
Results for Čebyšev systems of vector-valued functions are given by MOROZOV (1976) and ŠAŠKIN (1977).
Mairhuber-type theorems have been derived by DUNHAM (1968) for unisolvent families and by BRAESS (1972) for varisolvent families.

Chapter 3: Alternations were first used in the definition of Haar spaces by ZIELKE (1973), who also proved lemma 3.1. Another characterization of Haar spaces defined on finite sets M has been given by UHRIN (1977). The embedding theorems 3.3 and 3.4 are given in STOCKENBERG (1976). The proofs presented here are mostly based on work of ZALIK (1977a) who found theorem 3.3 independently.
Many examples of Haar spaces are found in KARLIN and STUDDEN (1966, p. 9-20), some more are contained in DUNHAM (1974) and ABAKUNOW (1972b). Textbooks on approximation theory (e.g. CHENEY 1966, LORENTZ 1966, SCHÖNHAGE 1971) usually deal with Haar spaces of continuous functions defined on intervals or on the unit circle. Taking lemma 3.1 c) as the definition of Čebyšev systems it is then easy

to derive zero and alternation properties via Rolle's theorem.

<u>Chapter 4:</u> Lemma 4.1 was shown by JONES and KARLOVITZ (1970) for intervals M and U c C(M). The general equivalence of a) and b) is due to BASTIEN and DUBUC (1976). In both papers the proofs are based on approximations by suitable Čebyšev systems. The direct argument given here is new. Most other results and proofs of this chapter follow (apart from minor modifications) STOCKENBERG (1977a) who based his work on a paper of BARTELT (1975). Theorem 4.5 and corollary 4.6 were independently and with different methods shown by SOMMER and STRAUSS (1977) for the special case U c C(M), M being an interval.

<u>Chapter 5:</u> The dimension property of corollary 5.2 is well known for Haar spaces U c $C(S^1)$ (see, e.g., LORENTZ 1966). See also lemma 2.4. Most results of this chapter seem to be literally new, but are straightforward generalizations of other concepts.

<u>Chapter 6:</u> Lemmas 6.1 and 6.3 are due to KURSHAN and GOPINATH (1977). For continuous functions defined on intervals most of theorem 6.5 was shown by KREIN (1951); see also KARLIN and STUDDEN (1966) and VOLKOV (1969). A geometric interpretation was pointed out by ABAKUNOV (1972a). The general version of theorem 6.5 given here is new, as well as the limitations (see chapter 10). Corollary 6.6 sharpens two results of GOPINATH and KURSHAN (1977a). These authors also show theorem 6.7.
If M c $\mathbb{R}$ is compact M contains its infimum and supremum. So theorem 6.5 disproves a result of MANDLER (1975) claiming that if M consists of more than n components, there is an n-dimensional Haar space U c C(M) that does not contain a positive function.

<u>Chapter 7:</u> Theorem 7.1 and corollary 7.2 are taken from ZIELKE (1973), 7.2 a) and c) were shown before by NEMETH (1966) under more special conditions. The geometric interpretation 7.2 d) was given independently by HADELER (1973) and ZIELKE (1971). According to RUTMAN (1965), theorem 7.7 was proved by KREIN for the case that M is an interval and U c C(M). KREIN's proof does not seem to exist in the literature. The first proof of KREIN's result was given by NEMETH (1969). Under the assumption that M has property (B) theorem 7.7 was shown by ZIELKE (1973) who also proved lemma 7.5. Theorem 7.6 is from STOCKENBERG (1977b) and was independently and with a different proof found by SOMMER and STRAUSS (1977) for the special case of continuous functions on an interval. The general version of theorem 7.7 given here is due to ZALIK (1977a), the proofs are new.

Chapter 8: The results are new exept for lemma 8.7 a) which is from ZIELKE (1974).

Chapter 9: The results mostly constitute new, though not very deep generalizations of ZIELKE (1972).

Chapter 10: Examples 10.1, 10.2, 10.3 are from ZIELKE (1975). AMIR and ZIEGLER (1976) state that the space U in example 10.4 contains no (n-1)-dimensional subspace, but give no proof. Example 10.5 and the parts in the other examples referring to theorem 6.5, are new. In constructing more counterexamples, the usefulness of exercise 3) seems to be limited to the case n = 3 (or n = 4 best; see ABAKUNOV 1972 b). HAVERKAMP's (1978) example was found independently of ZIELKE (1975).

Chapter 11: RUTMAN (1965) stated a more special version of theorem 11.3, bute gave no proof. All results are slight extensions of ZIELKE (1974). Lemma 11.1 together with lemma 8.7 yields a generalization of theorem 2 (a) in ZALIK (1977b).

Chapter 12: Lemma 12.2 and theorem 12.6 are stated without proof in RUTMAN (1965). ZALIK (1977b) claims a counterexample to lemma 12.2, but we think it is invalid. Lemmas 12.1 and 12.5 are with different proofs given by ZALIK (1977b, theorem 2(a) and (c)). The proofs in this chapter, as well as lemma 12.3, are new.

Chapter 13: Theorem 13.1 was found by BRASS (1977), the proof given here is ours. Corollary 13.4 is due to PASSOW (1973), the other results of the second application are taken from ZIELKE (1977).

Chapter 14: The argument follows ZIELKE (1974). The first results about adjoined functions seem to be from LAASONEN (1949).

Chapter 15: Our term "extended Markov system of order k" corresponds to the term "extended complete Tchebycheff-system of order k+1" in KARLIN and STUDDEN (1966, p. 6). The results of this chapter, as well as those of the next, are well known and can, for example, be found in KARLIN and STUDDEN (1966). See also VIDENSKII (1976).

Chapter 17: The results and general lines of the proofs are from ZALIK (1975), some details have been simplified. KREIN (1951) mentioned that RUTMAN has an example of a Haar space in C(M), M being an interval, to which no function can be adjoined. This example does not seem to have been published; its validity would contradict theorem 17.3.

Chapter 18: Our presentation is a streamlined version of ZALIK (1976 b). The author claims the existence of adjoined functions for the general case of periodic Haar spaces of continuous functions. However, the proof is based on an integral representation given by RUTMAN (1965) without a proof. ZALIK (1976) refers to a proof of RUTMAN's representation in another paper (ZALIK 1977b), but in that paper he actually disproves RUTMAN's representation and shows a more special result which, in our opinion, does not yield the existence of adjoined functions in the general case. We have not included ZALIK's integral representation in this book because his proof is very technical and too long to be reproduced here in sufficient detail. It might be worth while to find a more transparent argument for this interesting result.

Chapter 19: Our results include a number of more special statements on convex functions; see, e.g., chapter XI in KARLIN and STUDDEN (1966), MÜHLBACH (1973), ZALIK (1976a).

Chapter 20: The material is based on KURSHAN and GOPINATH (1977), our proof may be a little simpler.

Chapter 21: Most of the results of this chapter, especially theorem 21.6, are from GOPINATH and KURSHAN (1977b), where historical roots are discussed in extenso. Our presentation may be simpler than the original proof.

Chapter 22: Theorems 22.1 to 22.6 are well known. Apart from the "converse" statements b) => a) in theorems 22.7, 22.8 and 22.9, there exist a "converse" theorem involving t-oscillations by GOPINATH and KURSHAN (1977b), another "converse" involving strong unicity by MCLAUGHLIN and SOMERS (1975) and a "converse" involving one-sided approximation by DUNHAM (1975).

Chapter 23: Theorem 23.3 seems to be new, the proof of theorem 23.4 is new. Theorems 23.2, 23.4, 23.5 and 23.6 can be found in COPPEL's bool (1971) on disconjugacy. Besides, theorem 23.4 was shown by DUNHAM (1970). Theorem 23.7 was given by NEMETH (1974). If c and d are conjugate points of U and $U|_{[c,d]}$ is a Haar space U can be extended by a few isolated points at best. Results in this direction are given by NEMETH (1974, 1975). Using a result of ANDREEV (1969), in the latter paper the following extension of exercise 3 is shown:

$$\text{Let } Ly := \frac{d}{dx^2}\left(\frac{d}{dx^2} + 1^2\right) \ldots \left(\frac{d}{dx^2} + n^2\right),$$

so $U = \text{span}\{1, x, \sin(x)\ \cos(x), \ldots, \sin(nx), \cos(nx)\}$. Then 0 and 2π are disconjugate and $U|_{[0,2\pi]}$ is a Haar space that can be extended by at most $2n-1$ points.

It is not known if U has a Markov basis for $n \geq 2$.

Literature

Abakunow, J.G. [1972a]: The distribution of the zeros of polynomials in a Čebyšev system (Russian). p. 3-11 in:
A collection of articles in the constructive theory of functions and the extremal problems of functional analysis. Kalinin. gos. Univ., Kalinin.

Abakunow, J.G. [1972b]: Čebyšev systems of four functions. (Russian). p. 14-25 in:
A collection of articles in the constructive theory of functions and the extremal problems of functional analysis. Kalinin. gos. Univ., Kalinin.

Amir, D. and Z. Ziegler [1976]: Korovkin shadows and Korovkin systems in C(S)-spaces.
University of Wisconsin, Mathematical Research Center, Technical Summary Report Nr. 1638.

Andreev, V.I. [1969]: On Čebyšev systems which cannot be continued over the boundary. (Russian).
Učen. Zap. Kalinin. gos. ped. Inst. 29, 15-18.

Bartelt, M.W. [1975]: Weak Chebyshev sets and splines.
J. Appr. Th. 14, 30-37.

Bastien, R. and S. Dubuc [1976]: Systèmes faibles de Tchebycheff et polynomes de Bernstein.
Canad. J. Math. 28, 653-658.

Braess, D. [1972]: On topological properties of sets admitting varisolvent functions.
Proc. Amer. Math. Soc. 34, 453-456.

Brass, H. [1977]: Quadraturverfahren.
Vandenhoeck and Ruprecht Verlag, Göttingen.

Cheney, E.W. [1966]: Introduction to approximation theory.
McGraw-Hill Book Co., New York.

Coppel, W.A. [1971]: Disconjugacy.
Lecture Notes in Mathematics, vol. 220. Springer-Verlag Berlin.

Curtis, P.C.Jr. [1959]: N-parameter families and best approximation.
Pacific J. Math. 9, 1013-1027.

Deutsch, F., Nürnberger, G. and I. Singer [1978]: In preparation (announced in Nürnberger 1978).

Dunham, C.B. [1968]: Unisolvence on multidimensional spaces.
Canad. Math. Bull. 11, 469-474.

Dunham, C.B. [1970]: Extendabilily of T*sets.
Aequat. Math. 5, 1-2.

Dunham, C.B. [1974]: Families satisfying the Haar condition.
J. Appr. Th. 12, 291-298.

Dunham, C.B. [1975]: Uniqueness in one-sided linear Chebyshev approximation.
J. Appr. Th. 15, 275-277.

Gopinath, B. and Kurshan, R.P. [1977a]: The existence in T-spaces of functions with prescribed alternations.
J. Appr. Th. 21, 143-150.

Gopinath, B. and R.P. Kurshan [1977b]: The oscillation theorem for Tchebycheff spaces of bounded functions, and a converse.
J. Appr. Th. 21, 151-173.

Hadeler, K.P. [1973]: Remarks on Haar systems.
J. Appr. Th. 7, 59-62.

Haverkamp, R. [1978]: Zum Nullstellensatz von Krein für Haarsche Räume.
J. Appr. Th., 23, 104-107.

Henderson, G.W. and B.R. Ummel [1973]: The nonexistence of complex Haar systems on nonplanar locally connected spaces.
Proc. Amer. Math. Soc. 39, 640-641.

Jones, R.C. and L.A. Karlovitz [1970]: Equioscillation under non-uniqueness in the approximation of continuous functions.
J. Appr. Th. 3, 138-145.

Karlin, S. and W.J. Studden [1966]: Tchebycheff systems: With applications in analysis and statistics.
Interscience Publishers J. Wiley, New York.

Kiefer, J. and J. Wolfowitz [1965]: On a theorem of Hoel and Levine on extrapolation designs.
Ann. Math. Stat. 36, 1627-1665.

Krein, M.G. [1951]: The ideas of P.L. Čebyšev and A.A. Markov in the theory of limiting values of integrals and their further developments.
Transl. Amer. Math. Soc. 2, 1-122.

Kurshan, R.P. and B. Gopinath [1977]: Embedding an arbitrary function into a Tchebycheff space.
J. Appr. Th. 21, 126-142.

Laasonen, P. [1949]: Einige Sätze über Tchebyscheffsche Funktionensysteme.
Ann. Acad. Sci. Fenn. 52, 3-24.

Lorentz, G.G. [1966]: Approximation of functions.
Holt, Rinehart and Winston, New York.

Lutts, J.A. [1964]: Topological spaces which admit unisolvent systems.
Transactions Amer. Math. Soc. 111, 440-448.

Mairhuber, J.C. [1956]: On Haar's theorem concerning Chebyshev approximation problems having unique solutions.
Proc. Amer. Math. Soc. 7, 609-615.

Mandler, J. [1975]: Sur les systèmes de Tchebycheff.
J. Appr. Th. 14, 38-42.

McLaughlin, H.W. and K.B. Somers [1975]: Another characterization of Haar subspaces.
J. Appr. Th. 14, 93-102.

Morozov, É.N. [1976]: Chebyshev subspaces of vector-valued functions.
Math. Notes 19, 209-212.

Nemeth, A.B. [1966]: Transformations of the Chebyshev systems.
Mathematica (Cluj) 8, 315-333.

Nemeth, A.B. [1969]: About the extension of the domain of definition of the Chebyshev systems defined on intervals of the real axis.
Mathematica (Cluj) 11, 307-310.

Nemeth, A.B. [1974]: Conjugate point classification with application to Chebyshev systems.
Revue Anal. Num. Theor. Approx. 3, 73-78.

Nemeth, A.B. [1975]: A geometrical approach to conjugate point classification for linear differential equations.
Revue Anal. Num. Theor. Approx. 4, 137-152.

Nürnberger, G. [1977]: Schnitte für die metrische Projektion.
J. Appr. Th. 20, 196-220.

Nürnberger, G. [1978]: Continuous selections for the metric projection and alternation.
Bericht Nr. 39 des Institutes für Angewandte Mathematik der Universität Erlangen-Nürnberg.

Obreschkoff, N. [1966]: Verteilung und Berechnung der Nullstellen reeller Polynome.
VEB Deutscher Verlag der Wissenschaften, Berlin.

Overdeck, J.M. [1971]: On the nonexistence of complex Haar systems.
Bull. Amer. Math. Soc. 77, 737-740.

Passow, E. [1973]: Alternating parity of Tchebycheff systems.
J. Appr. 9, 295-298.

Polya, G. and G. Szegö [1972]: Problems and theorems in analysis I.
Springer, New York.

Rutman, M.A. [1965]: Integral representation of functions forming a Markov series.
Dokl. Akad. Nauk SSSR 164, 989-992.

Šaškin, J.A. [1977]: On sets admitting Čebyšev vector systems (Russian).
Mat. Zametki 21, 199-207.

Schmeißer, G. and H. Schirmeier [1976]: Praktische Mathematik.
Walter de Gruyter Verlag, Berlin und New York.

Schoenberg, I.J. and T.C. Yang [1961]: On the unicity of solutions of problems of best approximation.
Ann. Mat. Pura Appl. 54, 1-12.

Schönhage, A. [1971]: Approximationstheorie.
Walter de Gruyter Verlag, Berlin.

Sieklucki, K. [1958]: Topological properties of sets admitting the Tchebycheff systems.
Bull. Acad. Polon. Sci. Sér. Sci. Math. Phys. Astronom. 6, 603-606.

Sommer, M. and H. Strauß [1977]: Eigenschaften von schwach tschebyscheffschen Räumen.
J. Appr. Th. 21, 257-268.

Stockenberg, B. [1976]: Zur Struktur von Čebyšev- und schwachen Čebyšev-Räumen.
Dissertation Duisburg.

Stockenberg, B. [1977a]: On the number of zeros of functions in a weak Tchebyshev-Space.
Math. Z. 156, 49-57.

Stockenberg, B. [1977b]: Subspaces of weak and oriented Tchebyshev-spaces.
Manuscr. Math. 20, 401-407.

Uhrin, B. [1977]: A characterization of finite Chebyshev sequence in $\mathbb{R}^n$.
Lin. Alg. Appl. 18, 59-74.

Videnskii, V.S. [1976]: Tchebycheff systems differentiable with respect to a given sequence of functions (Russian).
Izvestija Akad. Nauk. Armjan. SSR, Mat. 11, 345-354.

Volkov, V.I. [1958]: Some properties of Chebyshev systems. (Russian).
Učen. Zap. Kalinin. gos. ped. inst. 26, 41-48.

Volkov, V.I. [1969]: A certain generalization of S.N. Bernstein's theorem (Russian).
Ucen. Zap. Kalinin. gos. ped. inst. 69, 32-38.

Zalik, R.A. [1975]: Existence of Tchebycheff extensions.
J. Math. Anal. Appl. 51, 68-75.

Zalik, R.A. [1967a]: Smoothness properties of generalized convex functions.
Proc. Amer. Math. Soc. 56, 118-120.

Zalik, R.A. [1976b]: Extension of periodic Tchebycheff systems.
J. Math. Anal. Appl. 56, 373-378.

Zalik, R.A. [1977a]: On transforming a Tchebycheff system into a complete Tchebycheff system.
J. Appr. Th. 20, 220-222.

Zalik, R.A. [1977b]: Integral representation of Tchebycheff systems.
Pacific J. Math. 68, 553-568.

Zielke, R. [1971]: Zur Struktur von Tschebyscheff-Systemen.
Dissertation, Konstanz.

Zielke, R. [1972]: A remark on periodic Tchebyshev systems.
Manuscr. Math. 7, 325-329.

Zielke, R. [1973]: On transforming a Tchebyshev-system into a Markov-system.
J. Appr. Th. 9, 357-366.

Zielke, R. [1974]: Alternation properties of Tchebyshev-systems and the existence of adjoined functions.
J. Appr. Th. 10, 172-184.

Zielke, R. [1975]: Tchebyshev systems that cannot be transformed into Markov systems.
Manuscr. Math. 17, 67-71.

Zielke, R. [1977]: Remarks on a paper of Passow.
J. Appr. Th. 20, 162-164.

Index

Vol. 551: Algebraic K-Theory, Evanston 1976. Proceedings. Edited by M. R. Stein. XI, 409 pages. 1976.

Vol. 552: C. G. Gibson, K. Wirthmüller, A. A. du Plessis and E. J. N. Looijenga. Topological Stability of Smooth Mappings. V, 155 pages. 1976.

Vol. 553: M. Petrich, Categories of Algebraic Systems. Vector and Projective Spaces, Semigroups, Rings and Lattices. VIII, 217 pages. 1976.

Vol. 554: J. D. H. Smith, Mal'cev Varieties. VIII, 158 pages. 1976.

Vol. 555: M. Ishida, The Genus Fields of Algebraic Number Fields. VII, 116 pages. 1976.

Vol. 556: Approximation Theory. Bonn 1976. Proceedings. Edited by R. Schaback and K. Scherer. VII, 466 pages. 1976.

Vol. 557: W. Iberkleid and T. Petrie, Smooth S^1 Manifolds. III, 163 pages. 1976.

Vol. 558: B. Weisfeiler, On Construction and Identification of Graphs. XIV, 237 pages. 1976.

Vol. 559: J.-P. Caubet, Le Mouvement Brownien Relativiste. IX, 212 pages. 1976.

Vol. 560: Combinatorial Mathematics, IV, Proceedings 1975. Edited by L. R. A. Casse and W. D. Wallis. VII, 249 pages. 1976.

Vol. 561: Function Theoretic Methods for Partial Differential Equations. Darmstadt 1976. Proceedings. Edited by V. E. Meister, N. Weck and W. L. Wendland. XVIII, 520 pages. 1976.

Vol. 562: R. W. Goodman, Nilpotent Lie Groups: Structure and Applications to Analysis. X, 210 pages. 1976.

Vol. 563: Séminaire de Théorie du Potentiel. Paris, No. 2. Proceedings 1975–1976. Edited by F. Hirsch and G. Mokobodzki. VI, 292 pages. 1976.

Vol. 564: Ordinary and Partial Differential Equations, Dundee 1976. Proceedings. Edited by W. N. Everitt and B. D. Sleeman. XVIII, 551 pages. 1976.

Vol. 565: Turbulence and Navier Stokes Equations. Proceedings 1975. Edited by R. Temam. IX, 194 pages. 1976.

Vol. 566: Empirical Distributions and Processes. Oberwolfach 1976. Proceedings. Edited by P. Gaenssler and P. Révész. VII, 146 pages. 1976.

Vol. 567: Séminaire Bourbaki vol. 1975/76. Exposés 471–488. IV, 303 pages. 1977.

Vol. 568: R. E. Gaines and J. L. Mawhin, Coincidence Degree, and Nonlinear Differential Equations. V, 262 pages. 1977.

Vol. 569: Cohomologie Etale SGA $4\frac{1}{2}$. Séminaire de Géométrie Algébrique du Bois-Marie. Edité par P. Deligne. V, 312 pages. 1977.

Vol. 570: Differential Geometrical Methods in Mathematical Physics, Bonn 1975. Proceedings. Edited by K. Bleuler and A. Reetz. VIII, 576 pages. 1977.

Vol. 571: Constructive Theory of Functions of Several Variables, Oberwolfach 1976. Proceedings. Edited by W. Schempp and K. Zeller. VI, 290 pages. 1977

Vol. 572: Sparse Matrix Techniques, Copenhagen 1976. Edited by V. A. Barker. V, 184 pages. 1977.

Vol. 573: Group Theory, Canberra 1975. Proceedings. Edited by R. A. Bryce, J. Cossey and M. F. Newman. VII, 146 pages. 1977.

Vol. 574: J. Moldestad, Computations in Higher Types. IV, 203 pages. 1977.

Vol. 575: K-Theory and Operator Algebras, Athens, Georgia 1975. Edited by B. B. Morrel and I. M. Singer. VI, 191 pages. 1977.

Vol. 576: V. S. Varadarajan, Harmonic Analysis on Real Reductive Groups. VI, 521 pages. 1977.

Vol. 577: J. P. May, E_∞ Ring Spaces and E_∞ Ring Spectra. IV, 268 pages. 1977.

Vol. 578: Séminaire Pierre Lelong (Analyse) Année 1975/76. Edité par P. Lelong. VI, 327 pages. 1977.

Vol. 579: Combinatoire et Représentation du Groupe Symétrique, Strasbourg 1976. Proceedings 1976. Edité par D. Foata. IV, 339 pages. 1977.

Vol. 580: C. Castaing and M. Valadier, Convex Analysis and Measurable Multifunctions. VIII, 278 pages. 1977.

Vol. 581: Séminaire de Probabilités XI, Université de Strasbourg. Proceedings 1975/1976. Edité par C. Dellacherie, P. A. Meyer et M. Weil. VI, 574 pages. 1977.

Vol. 582: J. M. G. Fell, Induced Representations and Banach *-Algebraic Bundles. IV, 349 pages. 1977.

Vol. 583: W. Hirsch, C. C. Pugh and M. Shub, Invariant Manifolds. IV, 149 pages. 1977.

Vol. 584: C. Brezinski, Accélération de la Convergence en Analyse Numérique. IV, 313 pages. 1977.

Vol. 585: T. A. Springer, Invariant Theory. VI, 112 pages. 1977.

Vol. 586: Séminaire d'Algèbre Paul Dubreil, Paris 1975–1976 (29ème Année). Edited by M. P. Malliavin. VI, 188 pages. 1977.

Vol. 587: Non-Commutative Harmonic Analysis. Proceedings 1976. Edited by J. Carmona and M. Vergne. IV, 240 pages. 1977.

Vol. 588: P. Molino, Théorie des G-Structures: Le Problème d'Equivalence. VI, 163 pages. 1977.

Vol. 589: Cohomologie l-adique et Fonctions L. Séminaire de Géométrie Algébrique du Bois-Marie 1965–66, SGA 5. Edité par L. Illusie. XII, 484 pages. 1977.

Vol. 590: H. Matsumoto, Analyse Harmonique dans les Systèmes de Tits Bornologiques de Type Affine. IV, 219 pages. 1977.

Vol. 591: G. A. Anderson, Surgery with Coefficients. VIII, 157 pages. 1977.

Vol. 592: D. Voigt, Induzierte Darstellungen in der Theorie der endlichen, algebraischen Gruppen. V, 413 Seiten. 1977.

Vol. 593: K. Barbey and H. König, Abstract Analytic Function Theory and Hardy Algebras. VIII, 260 pages. 1977.

Vol. 594: Singular Perturbations and Boundary Layer Theory, Lyon 1976. Edited by C. M. Brauner, B. Gay, and J. Mathieu. VIII, 539 pages. 1977.

Vol. 595: W. Hazod, Stetige Faltungshalbgruppen von Wahrscheinlichkeitsmaßen und erzeugende Distributionen. XIII, 157 Seiten. 1977.

Vol. 596: K. Deimling, Ordinary Differential Equations in Banach Spaces. VI, 137 pages. 1977.

Vol. 597: Geometry and Topology, Rio de Janeiro, July 1976. Proceedings. Edited by J. Palis and M. do Carmo. VI, 866 pages. 1977.

Vol. 598: J. Hoffmann-Jørgensen, T. M. Liggett et J. Neveu, Ecole d'Eté de Probabilités de Saint-Flour VI – 1976. Edité par P.-L. Hennequin. XII, 447 pages. 1977.

Vol. 599: Complex Analysis, Kentucky 1976. Proceedings. Edited by J. D. Buckholtz and T. J. Suffridge. X, 159 pages. 1977.

Vol. 600: W. Stoll, Value Distribution on Parabolic Spaces. VIII, 216 pages. 1977.

Vol. 601: Modular Functions of one Variable V, Bonn 1976. Proceedings. Edited by J.-P. Serre and D. B. Zagier. VI, 294 pages. 1977.

Vol. 602: J. P. Brezin, Harmonic Analysis on Compact Solvmanifolds. VIII, 179 pages. 1977.

Vol. 603: B. Moishezon, Complex Surfaces and Connected Sums of Complex Projective Planes. IV, 234 pages. 1977.

Vol. 604: Banach Spaces of Analytic Functions, Kent, Ohio 1976. Proceedings. Edited by J. Baker, C. Cleaver and Joseph Diestel. VI, 141 pages. 1977.

Vol. 605: Sario et al., Classification Theory of Riemannian Manifolds. XX, 498 pages. 1977.

Vol. 606: Mathematical Aspects of Finite Element Methods. Proceedings 1975. Edited by I. Galligani and E. Magenes. VI, 362 pages. 1977.

Vol. 607: M. Métivier, Reelle und Vektorwertige Quasimartingale und die Theorie der Stochastischen Integration. X, 310 Seiten. 1977.

Vol. 608: Bigard et al., Groupes et Anneaux Réticulés. XIV, 334 pages. 1977.

Vol. 609: General Topology and Its Relations to Modern Analysis and Algebra IV. Proceedings 1976. Edited by J. Novák. XVIII, 225 pages. 1977.

Vol. 610: G. Jensen, Higher Order Contact of Submanifolds of Homogeneous Spaces. XII, 154 pages. 1977.

Vol. 611: M. Makkai and G. E. Reyes, First Order Categorical Logic. VIII, 301 pages. 1977.

Vol. 612: E. M. Kleinberg, Infinitary Combinatorics and the Axiom of Determinateness. VIII, 150 pages. 1977.

Vol. 613: E. Behrends et al., L^P-Structure in Real Banach Spaces. X, 108 pages. 1977.

Vol. 614: H. Yanagihara, Theory of Hopf Algebras Attached to Group Schemes. VIII, 308 pages. 1977.

Vol. 615: Turbulence Seminar, Proceedings 1976/77. Edited by P. Bernard and T. Ratiu. VI, 155 pages. 1977.

Vol. 616: Abelian Group Theory, 2nd New Mexico State University Conference, 1976. Proceedings. Edited by D. Arnold, R. Hunter and E. Walker. X, 423 pages. 1977.

Vol. 617: K. J. Devlin, The Axiom of Constructibility: A Guide for the Mathematician. VIII, 96 pages. 1977.

Vol. 618: I. I. Hirschman, Jr. and D. E. Hughes, Extreme Eigen Values of Toeplitz Operators. VI, 145 pages. 1977.

Vol. 619: Set Theory and Hierarchy Theory V, Bierutowice 1976. Edited by A. Lachlan, M. Srebrny, and A. Zarach. VIII, 358 pages. 1977.

Vol. 620: H. Popp, Moduli Theory and Classification Theory of Algebraic Varieties. VIII, 189 pages. 1977.

Vol. 621: Kauffman et al., The Deficiency Index Problem. VI, 112 pages. 1977.

Vol. 622: Combinatorial Mathematics V, Melbourne 1976. Proceedings. Edited by C. Little. VIII, 213 pages. 1977.

Vol. 623: I. Erdelyi and R. Lange, Spectral Decompositions on Banach Spaces. VIII, 122 pages. 1977.

Vol. 624: Y. Guivarc'h et al., Marches Aléatoires sur les Groupes de Lie. VIII, 292 pages. 1977.

Vol. 625: J. P. Alexander et al., Odd Order Group Actions and Witt Classification of Innerproducts. IV, 202 pages. 1977.

Vol. 626: Number Theory Day, New York 1976. Proceedings. Edited by M. B. Nathanson. VI, 241 pages. 1977.

Vol. 627: Modular Functions of One Variable VI, Bonn 1976. Proceedings. Edited by J.-P. Serre and D. B. Zagier. VI, 339 pages. 1977.

Vol. 628: H. J. Baues, Obstruction Theory on the Homotopy Classification of Maps. XII, 387 pages. 1977.

Vol. 629: W.A. Coppel, Dichotomies in Stability Theory. VI, 98 pages. 1978.

Vol. 630: Numerical Analysis, Proceedings, Biennial Conference, Dundee 1977. Edited by G. A. Watson. XII, 199 pages. 1978.

Vol. 631: Numerical Treatment of Differential Equations. Proceedings 1976. Edited by R. Bulirsch, R. D. Grigorieff, and J. Schröder. X, 219 pages. 1978.

Vol. 632: J.-F. Boutot, Schéma de Picard Local. X, 165 pages. 1978.

Vol. 633: N. R. Coleff and M. E. Herrera, Les Courants Résiduels Associés à une Forme Méromorphe. X, 211 pages. 1978.

Vol. 634: H. Kurke et al., Die Approximationseigenschaft lokaler Ringe. IV, 204 Seiten. 1978.

Vol. 635: T. Y. Lam, Serre's Conjecture. XVI, 227 pages. 1978.

Vol. 636: Journées de Statistique des Processus Stochastiques, Grenoble 1977, Proceedings. Edité par Didier Dacunha-Castelle et Bernard Van Cutsem. VII, 202 pages. 1978.

Vol. 637: W. B. Jurkat, Meromorphe Differentialgleichungen. VII, 194 Seiten. 1978.

Vol. 638: P. Shanahan, The Atiyah-Singer Index Theorem, An Introduction. V, 224 pages. 1978.

Vol. 639: N. Adasch et al., Topological Vector Spaces. V, 125 pages. 1978.

Vol. 640: J. L. Dupont, Curvature and Characteristic Classes. X, 175 pages. 1978.

Vol. 641: Séminaire d'Algèbre Paul Dubreil, Proceedings Paris 1976–1977. Edité par M. P. Malliavin. IV, 367 pages. 1978.

Vol. 642: Theory and Applications of Graphs, Proceedings, Michigan 1976. Edited by Y. Alavi and D. R. Lick. XIV, 635 pages. 1978.

Vol. 643: M. Davis, Multiaxial Actions on Manifolds. VI, 141 pages. 1978.

Vol. 644: Vector Space Measures and Applications I, Proceedings 1977. Edited by R. M. Aron and S. Dineen. VIII, 451 pages. 1978.

Vol. 645: Vector Space Measures and Applications II, Proceedings 1977. Edited by R. M. Aron and S. Dineen. VIII, 218 pages. 1978.

Vol. 646: O. Tammi, Extremum Problems for Bounded Univalent Functions. VIII, 313 pages. 1978.

Vol. 647: L. J. Ratliff, Jr., Chain Conjectures in Ring Theory. VIII, 133 pages. 1978.

Vol. 648: Nonlinear Partial Differential Equations and Applications, Proceedings, Indiana 1976–1977. Edited by J. M. Chadam. VI, 206 pages. 1978.

Vol. 649: Séminaire de Probabilités XII, Proceedings, Strasbourg, 1976–1977. Edité par C. Dellacherie, P. A. Meyer et M. Weil. VIII, 805 pages. 1978.

Vol. 650: C*-Algebras and Applications to Physics. Proceedings 1977. Edited by H. Araki and R. V. Kadison. V, 192 pages. 1978.

Vol. 651: P. W. Michor, Functors and Categories of Banach Spaces. VI, 99 pages. 1978.

Vol. 652: Differential Topology, Foliations and Gelfand-Fuks-Cohomology, Proceedings 1976. Edited by P. A. Schweitzer. XIV, 252 pages. 1978.

Vol. 653: Locally Interacting Systems and Their Application in Biology. Proceedings, 1976. Edited by R. L. Dobrushin, V. I. Kryukov and A. L. Toom. XI, 202 pages. 1978.

Vol. 654: J. P. Buhler, Icosahedral Golois Representations. III, 143 pages. 1978.

Vol. 655: R. Baeza, Quadratic Forms Over Semilocal Rings. VI, 199 pages. 1978.

Vol. 656: Probability Theory on Vector Spaces. Proceedings, 1977. Edited by A. Weron. VIII, 274 pages. 1978.

Vol. 657: Geometric Applications of Homotopy Theory I, Proceedings 1977. Edited by M. G. Barratt and M. E. Mahowald. VIII, 459 pages. 1978.

Vol. 658: Geometric Applications of Homotopy Theory II, Proceedings 1977. Edited by M. G. Barratt and M. E. Mahowald. VIII, 487 pages. 1978.

Vol. 659: Bruckner, Differentiation of Real Functions. X, 247 pages. 1978.

Vol. 660: Equations aux Dérivée Partielles. Proceedings, 1977. Edité par Pham The Lai. VI, 216 pages. 1978.

Vol. 661: P. T. Johnstone, R. Paré, R. D. Rosebrugh, D. Schumacher, R. J. Wood, and G. C. Wraith, Indexed Categories and Their Applications. VII, 260 pages. 1978.

Vol. 662: Akin, The Metric Theory of Banach Manifolds. XIX, 306 pages. 1978.

Vol. 663: J. F. Berglund, H. D. Junghenn, P. Milnes, Compact Right Topological Semigroups and Generalizations of Almost Periodicity. X, 243 pages. 1978.

Vol. 664: Algebraic and Geometric Topology, Proceedings, 1977. Edited by K. C. Millett. XI, 240 pages. 1978.

Vol. 665: Journées d'Analyse Non Linéaire. Proceedings, 1977. Edité par P. Bénilan et J. Robert. VIII, 256 pages. 1978.

Vol. 666: B. Beauzamy, Espaces d'Interpolation Réels: Topologie et Géometrie. X, 104 pages. 1978.

Vol. 667: J. Gilewicz, Approximants de Padé. XIV, 511 pages. 1978.

Vol. 668: The Structure of Attractors in Dynamical Systems. Proceedings, 1977. Edited by J. C. Martin, N. G. Markley and W. Perrizo. VI, 264 pages. 1978.

Vol. 669: Higher Set Theory. Proceedings, 1977. Edited by G. H. Müller and D. S. Scott. XII, 476 pages. 1978.